Real Harmonic Analysis

**ANU
eVIEW**

Published by ANU eView
The Australian National University
Canberra ACT 0200, Australia
Email: anupress@anu.edu.au
This title is also available online at press.anu.edu.au

National Library of Australia Cataloguing-in-Publication entry

Author: Auscher, Pascal.

Title: Real harmonic analysis / lectures by Pascal Auscher with the assistance of
                         Lashi Bandara.

ISBN: 9781921934076 (pbk.) 9781921934087 (ebook)

Subjects: Mathematical analysis.
                         Mathematics.

Other Authors/Contributors:
                         Bandara, Lashi.

Dewey Number: 510

Real Harmonic Analysis

Lectures by Pascal Auscher
with the assistance of Lashi Bandara

Australian
National
University

eVIEW

Contents

Preface

Pascal Auscher recently spent a year at ANU on leave from Université Paris-Sud, where he is Professor of Mathematics. During this time he taught this graduate level course covering fundamental topics in modern harmonic analysis.

Auscher has made substantial contributions to harmonic analysis and partial differential equations, in a wide range of areas including functional calculi of operators, heat kernel estimates, Hardy spaces, weighted norm estimates and boundary value problems.

In particular his contributions were essential to the recent solution of a long-standing conjecture known as the Kato square root problem. This involved substantial new developments concerning the so-called Tb theorems and their application to singular integral operators or more generally to the functional calculus of operators which satisfy Davies-Gaffney estimates. These concepts were fundamental in the solution of the Kato problem on square roots of elliptic operators by Auscher, Hofmann, Lacey, Tchamitchian and myself.

Auscher has two published books to his name: *Square Root problem for divergence operators and related topics*, (with Ph. Tchamitchian) Astérisque vol 249, Soc. Math. de France 1998; and *On necessary and sufficient conditions for L^p estimates of Riesz transforms associated to elliptic operators on R^n and related estimates*, Memoirs of the American Math. Society vol. 186, Am. Math. Soc. 2007.

In this lecture course, Auscher provides a basic grounding in the advanced mathematics required for tackling problems in modern harmonic analysis and applying them, for example to partial differential equations. He does this from the point of view of an expert who fully understands the significance of the more basic material. Throughout the manuscript the material is developed in novel ways including some original proofs, derived from the advanced outlook of the author.

Thus the book is a mixture of expository material developed from a contemporary perspective and new work. It is likely to serve two audiences, both graduate students, and established researchers wishing to familiarise themselves with modern techniques of harmonic analysis. The lecture notes were taken and written up by an ANU PhD student, Lashi Bandara, thus providing the final polished account of the course material.

Alan McIntosh, Professor of Mathematics, MSI, ANU

Introduction

This book presents the material covered in graduate lectures delivered at the Australian National University in 2010. This material had also been partially presented at the University Paris-Sud before but never published.

Real Harmonic Analysis originates from the seminal works of Zygmund and Calderón, pursued by Stein, Weiss, Fefferman, Coifman, Meyer and many others.

Moving from the classical periodic setting to the real line, then to higher dimensional Euclidean spaces and finally to, nowadays, sets with minimal structures, the theory has reached a high level of applicability. This is why it is called real harmonic analysis: the usual exponential functions have disappeared from the picture. Set and function decomposition prevail.

This development has serve to solve famous conjectures in the field. The first one is the boundedness of the Cauchy integral operator on Lipschitz curves by Coifman, McIntosh and Meyer thirthy years ago. In the last ten years, there has been the solution of the Kato conjecture at the intersection of partial differential equations, functional analysis and harmonic analysis, and of the Painlevé conjecture at the border of complex function theory and geometric analysis. We mention also the breakthroughs on weighted norm inequalities with sharp behavior on the weight constants that are occurring at the moment, many of the articles being still in submitted form.

Nowadays all these developments bear on variations of boundedness criteria for singular integral operators and quadratic functionals called Tb theorems after the pionering work of McIntosh and Meyer followed by David, Journé and Semmes. These are powerful and versatile tools. However, they represent more advanced topics than such lectures could cover: these would be a follow-up topic. We have chosen to prepare the grounds to more advanced reading by presenting the basic material that is considered as "well-known" by experts. Nevertheless, anyone who wants to explore this field or apply it to some other problem must know the material we have chosen here. These lecture notes are therefore introductory to the field and accessible to beginners. They do not pretend to cover everything but a selection of important topics. Even if some lecture notes like this one exist, there is a need for updates as they do not cover the same material or address different points.

Let us present it in some more detail. We cover ten chapters: Measure theory, Coverings and cubes, Maximal functions, Interpolation, Bounded Mean Oscillation, Hardy spaces,

Calderón-Zygmund Operators, Carleson measures and BMO, Littlewood-Paley estimates and the T1 theorem for Singular Integrals.

The first chapter is to recall some notation and results from integration theory. The second chapter presents the fundamental tools of modern real harmonic analysis: covering lemmas of various sorts (Vitali, Besicovitch, Whitney). We have chosen to mostly restrict our setting to the Euclidean space to emphasize the ideas rather than the technicalities, Except for the Besicovitch covering lemma, of which we give a proof, they all extend to very general settings and we have tried to give proofs that are not too different from the ones in the extended settings. Maximal functions are treated in Chapter 3. These central objects can be declined in several forms: centred, uncentred, dyadic. They all serve the same goal: decompose the space into sets of controlled sizes according to a function. Chapter 4 on interpolation is rather standard and we present a simple as possible treatment. Chapter 5 on bounded mean oscillation restricts its attention to the main properties of BMO functions, the John-Nirenberg inequality and the technique of good lambda inequalities to prove the sharp function inequality of Fefferman-Stein. We also cover the recently developed good lambda inequality with two parameters which is useful for proving inequalities in restricted ranges of the exponents of the Lebesgue spaces. Chapter 6 presents atomic Hardy spaces: Coifman-Weiss' equivalence of the atomic definitions with p-atoms with $1 < p \leq \infty$ as a consequence of Calderón-Zygmund decompositions on functions, and applications to the Fefferman duality theorem. Chapter 7 is also an introduction to the theory of Calderón-Zygmund operators and some applications to multipliers are given towards Littlewood-Paley theorem. Again more advanced results can be obtained form the references. Chapter 8 covers the notion of a Carleson measure and Carleson's embedding theorem is proved. Then the connection between Carleson measures and BMO functions is studied in detail with care on the subtle convergence issues that are not often treated in the literature. Applications to the Bony-Coifman-Meyer paraproducts are given. Chapter 9 on Littlewood-Paley estimates could also be called T1 theorem for quadratic functionals and is extracted with details from an important article of Christ-Journé. See also my previous book with Philippe Tchamitchian were not all details are given. Almost orthogonality arguments, in particular the Cotlar-Knapp-Stein lemma, are central here. We finish in Chapter 1 with a proof of the T1 theorem of David-Journé. This proof is not the original one but is the one given by Coifman-Meyer and deserves more publicity. Again the strategy of this proof adapts to very general settings.

Special thanks go to Alan McIntosh and to the Mathematical Science Institute by inviting me as a one year visiting fellow at the Australian National University and also to the CNRS for giving me a "délégation" allowing a leave of absence from my home university. A one year visit that all my family has enjoyed and will remember for years.

Alan McIntosh and I have had and still have a long standing collaboration and it is a pleasure to thank him for sharing so many views on mathematics and life. That he accepted to preface this manuscript is a great privilege.

Thanks also go to Lashi Bandara whose typesetting genius made scratch notes he took from the lectures a readable manuscript. He also corrected some mistakes in the proofs (the remaining ones are my responsibility). My former student Frédéric Bernicot read the proofs entirely and found some more typos. I am grateful for that.

Addition to the second version: These lectures have been used again by myself at Université Paris-Sud and also by Thierry Coulhon and Dorothee Frey at ANU. I want to thank them for pointing out a number of remaining typos.

Addition to the third version (2022) : some further typos were eliminated after another class at university Paris-Saclay. Some chapters were slightly upgraded: Chapter 2 with the notion of quasi-metric space introduced.

Chapter 1

Measure Theory

While we shall focus our attention primarily on $\mathbb{R}^n$, we note some facts about measures in an abstract setting and in the absence of proofs.

Let X be a set. The reader will recall that in the literature, a measure μ on X is usually defined on a σ-algebra $\mathscr{M} \subset \mathscr{P}(X)$. This approach is limited in the direction we will take. In the sequel, it will be convenient to forget about measurability and associate a size to arbitrary subsets but still in a meaningful way. We present the *Carathéodory's* notion of measure and measurability. In the literature, what we call a measure is *sometimes* distinguished as an *outer measure*.

Definition 1.0.1 (Measure/Outer measure). *Let X be a set. Then, a map $\mu : \mathscr{P}(X) \to [0, +\infty]$ is called a measure on X if it satisfies:*

  (i) $\mu(\varnothing) = 0$,

  (ii) $\mu(A) \leq \sum_{i=1}^{\infty} \mu(A_i)$ whenever $A \subset \bigcup_{i=1}^{\infty} A_i$.

Remark 1.0.2. *Note that (ii) includes the statement: if $A \subset B$, then $\mu(A) \leq \mu(B)$.*

Certainly, the *classical* measures are additive on disjoint subsets. This is something we lose in the definition above. However, we can recover both the *measure σ-algebra* and a notion of measurable set.

Definition 1.0.3 (Measurable). *Let μ be a measure on X (in the sense of Definition 1.0.1). We say $A \subset X$ is μ-measurable if for all $Y \subset X$,*

$$\mu(A) = \mu(A \setminus Y) + \mu(A \cap Y).$$

Theorem 1.0.4. *Let $\{A_i\}_{i=1}^{\infty}$ be a countable set of μ-measurable sets. Then,*

  (i) $\bigcap_{i=1}^{\infty} A_i$ and $\bigcup_{i=1}^{\infty} A_i$ are μ-measurable,

  (ii) If the sets $\{A_i\}_{i=1}^{\infty}$ are mutually disjoint, then

$$\mu\left(\bigcup_{i=1}^{\infty} A_i\right) = \sum_{i=1}^{\infty} \mu(A_i),$$

(iii) If $A_k \subset A_{k+1}$ for all $k \geq 1$, then

$$\lim_{k \to \infty} \mu(A_k) = \mu\left(\bigcup_{i=1}^{\infty} A_i\right),$$

(iv) If $A_k \supset A_{k+1}$ for all $k \geq 1$, then

$$\lim_{k \to \infty} \mu(A_k) = \mu\left(\bigcap_{i=1}^{\infty} A_i\right),$$

(v) The set $\mathscr{M} = \{A \subset X : A \text{ is } \mu - \text{measurable}\}$ is a σ-algebra.

A proof of (i) - (iv) can can be found in [GC91, p2]. Then, (v) is an easy consequence and we leave it as an exercise.

The following result illustrates that we can indeed think about classical measures in this framework. Recall that a *measure space* is a triple $(X, \mathscr{M}, \nu)$ where $\mathscr{M} \subset \mathscr{P}(X)$ is a σ-algebra and $\nu : \mathscr{M} \to [0, +\infty]$ is a measure in the *classical* sense. Measure spaces are also given as the tuple (X, ν) by which we mean $(X, \mathscr{M}, \nu)$ where $\mathscr{M}$ is the largest σ-algebra containing ν-measurable sets.

Theorem 1.0.5. *Let $(X, \mathscr{M}, \nu)$ be a measure space. Then, there exists a measure μ in the sense of Definition 1.0.1 on X such that $\mu = \nu$ on $\mathscr{M}$.*

See [G.81, §5.2, Theorem 3].

Of particular importance is the following construction of the Lebesgue measure in our sense.

Definition 1.0.6 (Lebesgue Measure). *Let $A \subset \mathbb{R}^n$. For a Euclidean ball B, denote the volume of the ball by $\mathrm{vol}\, B$. Define the Lebesgue measure $\mathscr{L}$:*

$$\mathscr{L}(A) = \inf\left\{\sum_{i=1}^{\infty} \mathrm{vol}\, B_i : A \subset \bigcup_{i=1}^{\infty} B_i \text{ and each } B_i \text{ is an open ball}\right\}.$$

This definition is justified by the following proposition.

Proposition 1.0.7. *Let $(\mathbb{R}^n, \mathscr{M}', \mathscr{L}')$ be the classical Lebesgue measure defined on the largest possible σ-algebra $\mathscr{M}'$ and let $\mathscr{M} = \{A \subset X : A \text{ is } \mathscr{L} - \text{measurable}\}$. Then,*

$$\mathscr{M}' = \mathscr{M} \text{ and } \mathscr{L}(A) = \mathscr{L}'(A)$$

for all $A \in \mathscr{M}'$.

For a more detailed treatment of abstract measure theory, see [GC91, §1] and [G.81, Ch.5].

Chapter 2

Coverings and cubes

We will consider the setting of $\mathbb{R}^n$ with the usual Euclidean norm $|\cdot|$ inducing the standard Euclidean metric $d_E(x,y) = |x - y|$. We note, however, that some of the material that we discuss here can be easily generalised to a more abstract setting.

To introduce some nomenclature, we denote the ball centred at x with radius $r > 0$ by $B(x,r)$. We are intentionally ambiguous as to whether the ball is open or closed. We will specify when this becomes important.

For a ball B, let $\operatorname{rad} B$ denote its radius. For $\lambda > 0$, we denote the ball with the same centre but radius $\lambda \operatorname{rad} B$ by λB.

This chapter is motivated by the following two questions.

(1) Suppose that $\Omega = \bigcup_{B \in \mathscr{B}} B$ where $\mathscr{B}$ is a family of balls. We wish to extract a subfamily $\mathscr{B}'$ of balls that do not overlap "too much" and still cover Ω.

(2) Given a set $\Omega \subset \mathbb{R}^n$, how can we select a cover of Ω with a given geometric structure.

2.1 Vitali and Besicovitch

Lemma 2.1.1 (Vitali Covering Lemma). *Let $\{B_\alpha\}_{\alpha \in I}$ be a family of balls in $\mathbb{R}^n$ and suppose that*

$$\sup_{\alpha \in I} \operatorname{rad} B_\alpha < \infty.$$

Then there exists a subset $I_0 \subset I$ such that

(i) $\{B_\alpha\}_{\alpha \in I_0}$ are mutually disjoint.

(ii) $\bigcup_{\alpha \in I} B_\alpha \subset \bigcup_{\alpha \in I_0} 5B_\alpha$.

Remark 2.1.2. *(i) The balls $\{B_\alpha\}_{\alpha \in I}$ can be open or closed.*

(ii) This statement only relies on the metric structure of $\mathbb{R}^n$ with the euclidean metric given by $|\cdot|$. It can be replaced by a metric space (E, d). As an example, we can set $(E, d) = (\mathbb{R}^n, d_\infty)$, where d_∞ is the infinity distance given bc by the infinity norm. This is equivalent to replacing balls by cubes. It can even be replaced by a quasi-metric space: the distance d is replaced by a quasi-distance: it is a positive symmetric function d on $E \times E$, that is definite $(d(x, y) = 0$ if and only if $x = y)$ and that satisfies a quasi-triangle inequality $d(x, y) \leq A_0(d(x, z) + d(z, y)))$ for some $A_0 \geq 1$. In this case, one can check that 5 is replaced by $5A_0^2$.

(iii) The condition $\sup_{\alpha \in I} \operatorname{rad} B_\alpha < \infty$ is necessary. A counterexample is the collection of balls in $\mathbb{R}^n$ $\{B_i = B(0, i) : i \in \mathbb{N}\}$.

Proof. Let $M = \sup_{\alpha \in I} \operatorname{rad} B_\alpha$. For $j \in \mathbb{N}$, define

$$I(j) = \left\{\alpha \in I : 2^{-j-1}M < \operatorname{rad} B_\alpha \leq 2^{-j}M\right\}.$$

We inductively extract maximal subsets of each $I(j)$. So, for $j = 0$, let $J(0)$ be a maximal subset of $I(0)$ such that $\{B_\alpha\}_{\alpha \in J(0)}$ are mutually disjoint. The existence of such a collection is guaranteed by Zorn's Lemma. Now, for $j = 1$, we extract a maximal $J(1) \subset I(1)$ such that $\{B_\alpha\}_{\alpha \in J(1)}$ mutually disjoint and also disjoint from $\{B_\alpha\}_{\alpha \in J(0)}$. Now, when $j = k$, we let $J(k) \subset I(k)$ be maximal such that $\{B_\alpha\}_{\alpha \in J(k)}$ mutually disjoint and disjoint from $\{B_\alpha\}_{\alpha \in \cup_{m=0}^{k-1} J(m)}$. We then let

$$I_0 = \bigcup_{k \in \mathbb{N}} J(k).$$

By construction, $\{B_\alpha\}_{\alpha \in I_0}$ are mutually disjoint. This proves (i).

To prove (ii), fix a ball $B_\alpha \in I$. We show that there exists a $\beta \in I_0$ such that $B_\alpha \subset 5B_\beta$. We have a $k \in \mathbb{N}$ such that $\alpha \in I(k)$. That is,

$$2^{-k-1}M < \operatorname{rad} B_\alpha \leq 2^{-k}M.$$

If $\alpha \in J(k)$, then we're done. So suppose not. By construction, this must mean that B_α must intersect a ball B_β for $\beta \in J(l)$ where $0 \leq l \leq k$. But we know that

$$\operatorname{rad} B_\beta \geq 2^{-l-1}M \geq 2^{-k-1}M \geq \frac{1}{2}\operatorname{rad} B_\alpha$$

and by the triangle inequality

$$d(x_\alpha, x_\beta) \leq \operatorname{rad} B_\alpha + \operatorname{rad} B_\beta.$$

Then, $\operatorname{rad} B_\alpha + d(x_\alpha, x_\beta) \leq 2\operatorname{rad} B_\alpha + \operatorname{rad} B_\beta \leq 5\operatorname{rad} B_\beta$ and so it follows that

$$B_\alpha = B(x_\alpha, \operatorname{rad} B_\alpha) \subset B(x_\beta, \operatorname{rad} B_\alpha + d(x_\alpha, x_\beta)) \subset B(x_\beta, 5\operatorname{rad} B_\beta) = 5B_\beta.$$

$\square$

Remark 2.1.3. *By replacing powers 2^{-j} in the proof by power λ^{-j} with $\lambda > 1$, we can replace 5 by any number larger than 3.*

Before we introduce our next covering theorem, we require a rigorous notion of a family of balls to not intersect "too much." First, we note that χ_X denotes the indicator function of B.

Definition 2.1.4 (Bounded Overlap). *A collection of balls $\mathscr{B}$ is said to have bounded overlap if there exists a constant $C \in \mathbb{N}$ such that*

$$\sum_{B \in \mathscr{B}} \chi_B \leq C.$$

Theorem 2.1.5 (Besicovitch Covering Theorem). *Let $E \subset \mathbb{R}^n$. For each $x \in E$, let $B(x)$ be a ball centred at x. Assume that E is bounded or that $\sup_{x \in E} \operatorname{rad} B(x) < \infty$. Then, there exists a countable set $E_0 \subset E$ and a constant $C(n) \in \mathbb{N}$ such that*

(i) $E \subset \bigcup_{x \in E_0} B(x)$.

(ii) $\sum_{x \in E_0} \chi_{B(x)} \leq C(n)$.

Remark 2.1.6. (i) *Here, (ii) means that the set $\{B(x)\}_{x \in E_0}$ forms a bounded covering of E with constant $C(n)$. This constant depends only on dimension. The bounded covering property tells us that the balls are "almost disjoint." In fact, we can organise $E_0 = E_1 \cup \ldots \cup E_N$ such that each set $\{B(x)\}_{x \in E_k}$ contain mutually disjoint balls. We will not prove this.*

(ii) *We can substitute cubes for balls.*

(iii) *For E unbounded, the condition $\sup_{x \in E} \operatorname{rad} B(x) < \infty$ is necessary.*

(iv) *This theorem is very special to $\mathbb{R}^n$. A counterexample is the Heisenberg group with appropriate distance, one can show such a statement (with centred balls) fails.*

(v) *The balls must be centred at each point in E. Otherwise, consider $E = [0, 1)$ and $B_i = [0, 1 - 2^i], i \geq 1$.*

Before we proceed to prove the theorem, we require the following Lemma.

Lemma 2.1.7. *For $y \in \mathbb{R}^n$, let $C_\varepsilon(y)$ be the sector with vertex y with aperture angle ε. Suppose that $0 < \varepsilon \leq \pi/6$. Then, for all $R > 0$, $x, z \in C_\varepsilon(y)$, if $|x - y|, |z - y| \leq R$ then $|x - z| \leq R$.*

Proof. Expand $|(x - y) + (y - z)|^2$ and observe that the angle between $x - y$ and $z - y$ has cosine larger than or equal to $1/2$. $\square$

Proof of Besicovitch Covering Theorem. Let $M = \sup_{x \in E} \operatorname{rad} B(x)$ and suppose that E is bounded and $M = \infty$. Then, fix an $x_0 \in E$ and there exists a ball $B(x_0, R)$ such that $B(x_0, R) \supset E$. Then, we're done by setting $E_0 = \{x_0\}$.

So, we suppose now that E is bounded and $M < \infty$. Define:

$$E(k) = \left\{ x \in E : 2^{-k-1} M < \operatorname{rad} B(x) \leq 2^{-k} M \right\}.$$

We select points x_j inductively from each $E(k)$ to construct a set $E'(k)$. So, fix an initial $x_{0,0} \in E(0)$ and select $x_{0,i} \in E(0)$ by requiring that $x_{0,i} \notin \bigcup_{l=0}^{i-1} B(x_{0,l})$. Then, for arbitrary $k > 0$, assume that $E'(0), \ldots, E'(k-1)$ have already been constructed and construct $E'(k)$ by selecting $x_{k,i}$ such that $x_{k,i} \notin \bigcup_{l=0}^{i-1} B(x_{k,l}) \cup_{m=1}^{k-1} \cup_{x_{m,i} \in E'(m)} B(x_{m,i})$. Each $E'(k)$ must be finite since by the boundedness of E and the definition of $E(k)$, this process must stop after finitely many selections in $E(k)$. Now, let $E_0 = \bigcup_{k \in \mathbb{N}} E'(k)$ equipped with the natural ordering. That is, $E_0 = \{x_1, x_2, \ldots\}$ and if $i < j$ then $x_j \notin B(x_i)$. This is the same as saying that x_i was selected before x_j.

We prove (i). Suppose $x \in E$ but $x \notin \bigcup_{x_i \in E_0} B(x_i)$. In particular this means that $x \in E'(k)$ for some k which is a contradiction.

Now, to show (ii), fix $y \in \mathbb{R}^n$, let $\varepsilon = \pi/6$ and let $C_\varepsilon(y)$ denote the sector with vertex y and aperture ε. Define

$$A_y = \{x_i \in E_0 : y \in B(x_i) \text{ and } x_i \in C_\varepsilon(y)\},$$

and let x_i be the first element in A_y. Take $x_j \in A_y$ with $j > i$. Then,

$$\begin{cases} x_i, x_j \in C_\varepsilon(y) \\ |x_i - y| < \operatorname{rad} B(x_i) \\ |x_j - y| < \operatorname{rad} B(x_j) \end{cases}$$

and so $|x_i - x_j| < \max\{\operatorname{rad} B(x_i), \operatorname{rad} B(x_j)\}$ by application of Lemma 2.1.7. But by the ordering on E_0, x_j was selected after x_i and so $x_j \notin B(x_i)$. Consequently $\operatorname{rad} B(x_i) \leq |x_j - x_i|$ which implies $\operatorname{rad} B(x_j) > \operatorname{rad} B(x_i)$.

Now, suppose that x_i was selected at generation k, so $2^{-k-1} M < \operatorname{rad} B(x_i) \leq 2^{-k} M$. Also, $2^{-k-1} M < \operatorname{rad} B(x_j) \leq 2^{-k} M$ for otherwise, x_j would be selected at a later generation $l > k$ which implies that $\operatorname{rad} B(x_i) > \operatorname{rad} B(x_j)$ which is a contradiction.

Also, note that $\{B(x_j, 2^{-k-2} M) : j \in A_y\}$ are mutually disjoint and are all contained in $B(x_i, 2^{-k+1} M)$. It follows that,

$$\sum_{j \in A_y} \mathscr{L}(B(x_j, 2^{-k-2} M)) \leq \mathscr{L}(B(x_i, 2^{-k+1} M)) = (2^3)^n \mathscr{L}(B(0, 2^{-k-2} M))$$

and

$$\operatorname{card} A_y = \sum_{j \in A_y} 1 = \sum_{j \in A_y} \frac{\mathscr{L}(B(x_j, 2^{-k-2} M))}{\mathscr{L}(B(0, 2^{-k-2} M))} \leq 2^{3n}.$$

As $\mathbb{R}^n$ can be covered by a finite number of sectors $C_\varepsilon(y)$ we are done.

We shall not give details for the case that E is unbounded, but it can be obtained from the previous case with some effort. We refer the reader to [GC91, p35]. $\qquad \square$

Corollary 2.1.8 (Sard's Theorem). *Let $f : \mathbb{R}^n \to \mathbb{R}^n$ and let*

$$A = \left\{ x \in \mathbb{R}^n : \liminf_{r \to 0} \frac{\mathscr{L}(f(B(x,r)))}{\mathscr{L}(B(x,r))} = 0 \right\}.$$

Then $\mathscr{L}(f(A)) = 0$. (On the numerator $\mathscr{L}$ is understood as the exterior Lebesgue measure on $\mathbb{R}^n$).

Proof. Firstly, we note that for all $x \in A$ and $\varepsilon > 0$, there exists an $r_x \in (0,1]$ such that

$$\mathscr{L}(f(B(x, r_x))) \leq \varepsilon \mathscr{L}(B(x, r_x)).$$

Let $A_0 \subset A$ be the set of centres given by Besicovitch applied to the set of balls $\{B_x = B(x, r_x)\}$ for which the above measure condition holds. Then, $f(A) \subset \bigcup_{x_i \in A_0} f(B(x_i))$ and by the subadditivity of $\mathscr{L}$,

$$
\begin{aligned}
\mathscr{L}(f(A)) &\leq \sum_{x_i \in A_0} \mathscr{L}(f(B_{x_i})) \\
&\leq \varepsilon \sum_{x_i \in A_0} \mathscr{L}(B_{x_i}) \\
&\leq \varepsilon \int_{A + B(0,1)} \sum \chi_{B_{x_i}}(y) \, d\mathscr{L}(y) \\
&\leq \varepsilon C(n) \mathscr{L}(A + B(0,1)).
\end{aligned}
$$

Now, if A is bounded, then $\mathscr{L}(A + B(0,1)) < \infty$ and we can obtain the conclusion by letting $\varepsilon \to 0$. Otherwise, we replace A by $A \cap B(0, k)$ to obtain $\mathscr{L}(f(A \cap B(0,k))) = 0$ and then by taking $k \to \infty$ we establish $\mathscr{L}(f(A))) = 0$. $\qquad\square$

It is easy to show that A contains the singular points of f, that is points x at which f is differentiable with vanishing differential $df(x) = 0$. That is for a differentiable map, almost all points are regular (i.e., non-singular).

2.2 Dyadic Cubes

We begin with the construction of dyadic cubes on $\mathbb{R}^n$.

Definition 2.2.1 (Dyadic Cubes). *Let* $[0,1)^n$ *be the* reference cube *and let* $j \in \mathbb{Z}$ *and* $k = (k_1, \ldots, k_n) \in \mathbb{Z}^n$. *Then define the* dyadic cube *of generation* j *with lower left corner* $2^{-j}k$

$$\mathcal{Q}_{j,k} = \left\{ x \in \mathbb{R}^n : 2^j x - k \in [0,1)^n \right\},$$

the set of generation j *dyadic cubes*

$$\mathscr{Q}_j = \left\{ \mathcal{Q}_{j,k} : k \in \mathbb{Z}^n \right\},$$

and the set of all cubes

$$\mathscr{Q} = \bigcup_{j \in \mathbb{Z}} \mathscr{Q}_j = \left\{ \mathcal{Q}_{j,k} : j \in \mathbb{Z} \text{ and } k \in \mathbb{Z}^n \right\}.$$

We define the length of a cube to be its side length $\ell(\mathcal{Q}_{j,k}) = 2^{-j}$.

Remark 2.2.2. *If we were to replace* $[0,1)^n$ *with* $R = \prod_{i=1}^n [a_i, a_i + \delta)$ *as a reference cube, then the dyadic cubes with respect to* R *are constructed via the* homothety $\varphi : \mathbb{R}^n \to \mathbb{R}^n$ *where* $\varphi([0,1)^n) = R$.

Theorem 2.2.3 (Properties of Dyadic Cubes).   *(i) For $j \in \mathbb{Z}$, $\ell(\mathcal{Q}_{j,k}) = 2^{-j}$, $\mathscr{L}(\mathcal{Q}_{j,k}) = 2^{-jn}$, and $\mathcal{Q}_j$ forms a partition of $\mathbb{R}^n$ for each $j \in \mathbb{Z}$.*

(ii) For all $j \in \mathbb{Z}$ and $k \in \mathbb{Z}^n$, there exists a unique $k' \in \mathbb{Z}^n$ such that $\mathcal{Q}_{j,k} \subset \mathcal{Q}_{j-1,k'}$. We set $\widehat{\mathcal{Q}_{j,k}} = \mathcal{Q}_{j-1,k'}$ and it is called the parent of $\mathcal{Q}_{j,k}$.

(iii) For all $j \in \mathbb{Z}$ and $k \in \mathbb{Z}^n$, the cubes $\mathcal{Q}_{j+1,k'}$ for which $\mathcal{Q}_{j,k}$ is the parent are called the children of $\mathcal{Q}_{j,k}$.

(iv) For all $x \in \mathbb{R}^n$, there exists a unique sequence of dyadic cubes $(\mathcal{Q}_{j,k_j})_{j\in\mathbb{Z}} \subset \mathcal{Q}$ such that $x \in \mathcal{Q}_{j,k_j}$ and $\mathcal{Q}_{j,k_j} \in \mathcal{Q}_j$ for each $j \in \mathbb{Z}$. Moreover, $\mathcal{Q}_{j+1,k_{j+1}}$ is a child of $\mathcal{Q}_{j,k}$, $\mathcal{Q}_{j,k_j}$ is the parent of $\mathcal{Q}_{j+1,k_{j+1}}$. Besides, $\cap \mathcal{Q}_{j,k_j} = \{x\}$.

(v) For all $\mathcal{Q}, \mathcal{Q}' \in \mathcal{Q}$, either $\mathcal{Q} \subset \mathcal{Q}'$ or $\mathcal{Q}' \subset \mathcal{Q}$ or $\mathcal{Q} \cap \mathcal{Q}' = \varnothing$.

(vi) Let $\mathscr{E} \subset \mathcal{Q}$ such that $\Omega = \bigcup_{\mathcal{Q} \in \mathscr{E}} \mathcal{Q}$ satisfies $\mathscr{L}(\Omega) < \infty$. Then, the collection $\mathscr{F} = \{\mathcal{Q} \in \mathcal{Q} : \mathcal{Q} \subset \Omega, \widehat{\mathcal{Q}} \not\subset \Omega\}$ is composed of mutually disjoint dyadic cubes and $\Omega = \bigcup_{\mathcal{Q}' \in \mathscr{F}} \mathcal{Q}'$.

Proof. We leave it to the reader to verify (i) - (v). We prove (vi).

We prove that $\mathscr{F} \neq \varnothing$. Let $\mathcal{Q} \in \mathscr{E}$, and denote the kth parent by $\widehat{\mathcal{Q}}^k$. It follows that $\mathscr{L}(\widehat{\mathcal{Q}}^k) = (2^n)^k \mathscr{L}(\mathcal{Q})$. As $\mathscr{L}(\Omega) < \infty$, $k_0 = \max\left\{ k : \widehat{\mathcal{Q}}^k \subset \Omega \right\}$ is well-defined and consequently, $\widehat{\mathcal{Q}}^{k_0} \in \mathscr{F}$.

Now, note that $\bigcup_{\mathcal{Q} \in \mathscr{E}} \mathcal{Q} \subset \bigcup_{\mathcal{Q}' \in \mathscr{F}} \mathcal{Q}'$ and by the construction of $\mathscr{F}$, $\bigcup_{\mathcal{Q}' \in \mathscr{F}} \mathcal{Q}' \subset \Omega$. But by hypothesis $\Omega = \bigcup_{\mathcal{Q} \in \mathscr{E}} \mathcal{Q}$ and so it follows that $\Omega = \bigcup_{\mathcal{Q}' \in \mathscr{F}} \mathcal{Q}'$.

To complete the proof, we prove that the cubes in $\mathscr{F}$ are mutually disjoint. Let $\mathcal{Q}', \mathcal{Q}'' \in \mathscr{F}$ and assume that $\mathcal{Q}' \neq \mathcal{Q}''$. This is exactly that $\mathcal{Q}' \not\subset \mathcal{Q}''$ and $\mathcal{Q}' \not\supset \mathcal{Q}''$ so by (v), $\mathcal{Q}' \cap \mathcal{Q}'' = \varnothing$. $\qquad\square$

Remark 2.2.4.   *(i) The way (vi) is formulated does not ensure that $\mathscr{F} \subset \mathscr{E}$. Take for example $\mathscr{E} = \{[0, 1/2), [1/2, 1)\}$ which gives $\mathscr{F} = \{[0, 1)\}$.*

(ii) The assumption $\mathscr{L}(\Omega) < \infty$ cannot be dropped in (vi). A counterexample is $[0, 2^k) \subset \mathbb{R}$ for $k \in \mathbb{N}$.

Corollary 2.2.5. *Suppose Ω is an open set with $\mathscr{L}(\Omega) < \infty$. Then there exists a unique collection of disjoint dyadic cubes maximal for inclusion in Ω such that Ω is the union of such cubes. These cubes are usually called the maximal dyadic cubes for the inclusion $Q \subset \Omega$.*

Proof. We let $\mathscr{E} = \{\mathcal{Q} \in \mathcal{Q} : \mathcal{Q} \subset \Omega\}$. If $x \in \Omega$, then as Ω is open one can check that at least one of the cubes $\mathcal{Q}_{j,k_j}$ in (iv) is contained in Ω. Thus, Ω is the union of the cubes in $\mathscr{E}$. Hence $\mathscr{F}$ provides a partition of Ω with dyadic cubes. Besides we have $\mathscr{F} \subset \mathscr{E}$. Indeed we have clearly $\mathscr{F} = \left\{ \mathcal{Q} \in \mathscr{E} : \widehat{\mathcal{Q}} \not\subset \Omega \right\}$.

As for uniqueness, suppose we have two such collections $(\mathcal{Q}_i)$ and $(\mathcal{Q}'_j)$. For each j, then $\mathcal{Q}'_j$ meets at least one $\mathcal{Q}_i$. One is thus contained in the other. If the inclusion is strict, then there is inclusion of the parent, which implies that this parent is contained in Ω, contradicting maximality. Thus, there is equality and the two collections are identical. $\square$

Remark 2.2.6. *The idea of selecting maximal dyadic cubes can be adapted with many conditions. We shall see that right away with the Whitney covering theorem.*

2.3   Whitney coverings

The Whitney covering theorems are an important tool in Harmonic Analysis. We initially give a dyadic version of Whitney. But first, we recall some terminology from the theory of metric spaces. Recall that the diameter of a set $E \subset X$ for a (quasi-)metric space (X, d) is given by

$$\operatorname{diam} E = \sup_{x,y \in E} d(x, y),$$

and the (quasi-)distance from E to another set $F \subset X$ is given by

$$\operatorname{dist}(E, F) = \inf_{x \in E, y \in F} d(x, y).$$

Then the distance from E to a point $x \in X$ is simply given by $\operatorname{dist}(x, E) = \operatorname{dist}(E, F)$ where $F = \{x\}$.

Theorem 2.3.1 (Whitney Covering Theorem for Dyadic cubes). *Let $O \subsetneq \mathbb{R}^n$ be open. Then, there exists a collection of Dyadic cubes $\mathscr{F} = \{\mathcal{Q}_i\}_{i \in I}$ such that*

(i)

$$\frac{1}{30} \operatorname{dist}(\mathcal{Q}_i, {}^c O) \leq \operatorname{diam} \mathcal{Q}_i \leq \frac{1}{10} \operatorname{dist}(\mathcal{Q}_i, {}^c O),$$

(ii)

$$O = \bigcup_{i \in I} \mathcal{Q}_i,$$

(iii) The dyadic cubes in $\mathscr{F}$ are mutually disjoint.

Proof. We define $\mathscr{E}$ as the collection of dyadic cubes $\mathcal{Q}$ such that:

(a) $\mathcal{Q} \subset O$,

(b) $\operatorname{diam} \mathcal{Q} \leq \frac{1}{10} \operatorname{dist}(\mathcal{Q}, {}^c O)$.

The collection $\mathscr{E}$ is not empty. Let $x \in O$ and $(\mathcal{Q}_{j,k_j})_{j \in \mathbb{Z}}$ the sequence of dyadic cubes which contains x. So, there exists a $\mathcal{Q}_{j,k_j} \subset O$ and since x is fixed, we can impose the condition (b) as ${}^c O$ is not empty.

As in Corollary 2.2.5, there is a (unique) collection $\mathscr{F}$ of maximal dyadic cubes of O for the conditions (a) and (b). Here, we do not need that O has finite measure: we use that in (b), the right hand side increases by taking parent while the right hand side decreases. This proves (ii), and (iii) and by construction $\operatorname{diam} Q \leq \frac{1}{10} \operatorname{dist}(Q, {}^cO)$ for every $Q \in \mathscr{F}$. It remains to check the lower bound.

So, take $Q \in \mathscr{F}$. Then, by maximality, either $\widehat{Q} \not\subset O$ or $\operatorname{dist}(\widehat{Q}, {}^cO) < 10 \operatorname{diam} \widehat{Q}$. Thus, in either case,

$$\operatorname{dist}(Q, {}^cO) \leq 11 \operatorname{diam} \widehat{Q}.$$

Combining this with the fact that $\operatorname{diam} \widehat{Q} = 2 \operatorname{diam} Q$, we find

$$\operatorname{dist}(Q, {}^cO) \leq 22 \operatorname{diam} Q \leq 30 \operatorname{diam} Q$$

and this completes the proof. □

Remark 2.3.2. *In (i), the constant $1/10$ could be replaced by $1/2 - \varepsilon$ for all $\varepsilon > 0$. However, this would change the constant $1/30$ in the lower bound.*

The Whitney dyadic cubes introduced in the preceding theorem satisfy some important properties.

Proposition 2.3.3. *The Whitney Dyadic cubes of O satisfy:*

(i) For all $i \in I$, $3Q_i \subset O$,

(ii) For all $i, j \in I$, if $3Q_i \cap 3Q_j \neq \varnothing$, then

$$\frac{1}{4} \leq \frac{\operatorname{diam} Q_i}{\operatorname{diam} Q_j} \leq 4,$$

(iii) There exists a constant only depending on dimension, $C(n)$, such that

$$\sum_{i \in I} \chi_{3Q_i} \leq C(n).$$

Proof. (i) Suppose there exists an $z \in {}^cO \cap 3Q_i$. So, there exists a $y \in Q_i$ such that $d(y, z) \leq \operatorname{diam} Q_i$. But,

$$10 \operatorname{diam} Q_i \leq \operatorname{dist}(Q_i, {}^cO) \leq \operatorname{dist}(y, {}^cO) \leq \operatorname{dist}(y, z) \leq \operatorname{diam} Q_i$$

which is a contradiction.

(ii) By symmetry, it suffices to prove that

$$\frac{\operatorname{diam} Q_i}{\operatorname{diam} Q_j} \leq 4.$$

Let $y \in 3Q_i \cap 3Q_j$. We note that $\operatorname{dist}(y, Q_i) \leq \operatorname{diam} Q_i$ since $y \in 3Q_i$, and by the triangle inequality,

$$10 \operatorname{diam} Q_i \leq \operatorname{dist}(Q_i, {}^cO) \leq \operatorname{dist}(y, {}^cO) + \operatorname{dist}(y, Q_i) \leq \operatorname{dist}(y, {}^cO) + \operatorname{diam} Q_i$$

which shows that $\operatorname{dist}(y, {}^cO) \geq 9 \operatorname{diam} Q_i$.

15

Also, there exists $z \in \mathcal{Q}_j$ such that $d(y, z) \le \operatorname{diam} \mathcal{Q}_j$ and $\operatorname{dist}(y, {}^c O) \le \operatorname{dist}(z, {}^c O) + \operatorname{diam} \mathcal{Q}_j$. We estimate $\operatorname{dist}(z, {}^c O)$. Fix $w \in \mathcal{Q}_j$ and we find

$$\operatorname{dist}(z, {}^c O) \le d(z, w) + \operatorname{dist}(w, {}^c O) \le \operatorname{diam} \mathcal{Q}_j + \operatorname{dist}(w, {}^c O).$$

By taking an infimum over all $w \in \mathcal{Q}_j$ and using the fact that $\operatorname{dist}(\mathcal{Q}_j, {}^c O) \le 30 \operatorname{diam} \mathcal{Q}_j$, we find that $\operatorname{dist}(z, {}^c O) \le 31 \operatorname{diam} \mathcal{Q}_j$ and consequently $\operatorname{dist}(y, {}^c O) \le 32 \operatorname{diam} \mathcal{Q}_j$.

Putting these estimates together, we get that

$$\frac{\operatorname{diam} \mathcal{Q}_i}{\operatorname{diam} \mathcal{Q}_j} \le \frac{32}{9} < 4.$$

(iii) Fix $i \in I$ and let $A_i = \{j \in I : 3\mathcal{Q}_i \cap 3\mathcal{Q}_j \ne \varnothing\}$. So, for any $y \in 3\mathcal{Q}_i \cap 3\mathcal{Q}_j$,

$$\begin{cases} \operatorname{dist}(y, \mathcal{Q}_i) \le \operatorname{diam} \mathcal{Q}_i, \\ \operatorname{dist}(y, \mathcal{Q}_j) \le \operatorname{diam} \mathcal{Q}_j, \\ \operatorname{dist}(\mathcal{Q}_i, \mathcal{Q}_j) \le \operatorname{diam} \mathcal{Q}_i + \operatorname{diam} \mathcal{Q}_j. \end{cases}$$

Let $\mathcal{K} = \{-2, -1, 0, 1, 2\}$. Then, for any $j \in A_i$, $\operatorname{diam} \mathcal{Q}_j = 2^k \operatorname{diam} \mathcal{Q}_i$ where $k \in \mathcal{K}$. Now for such a $k \in \mathcal{K}$, define $A_i^k = \{j \in A_i : \operatorname{diam} \mathcal{Q}_j = 2^k \operatorname{diam} \mathcal{Q}_i\}$.

If $j \in A_i^k$, then $\operatorname{dist}(\mathcal{Q}_i, \mathcal{Q}_j) \le (1 + 2^k) \operatorname{diam} \mathcal{Q}_i \le 5 \operatorname{diam} \mathcal{Q}_i$, and in particular, this means that $\mathcal{Q}_j \subset 10\mathcal{Q}_i$. But all such $\{\mathcal{Q}_j\}_{j \in A_i^k}$ are mutually disjoint and so $\mathcal{L}\left(\bigcup_{j \in A_i^k} \mathcal{Q}_j\right) \le (10)^n \mathcal{L}(\mathcal{Q}_i) = 40^n \mathcal{L}(\mathcal{Q}_j)$ for any $j \in A_i^k$ since $\operatorname{diam} \mathcal{Q}_i = 2^{-k} \operatorname{diam} \mathcal{Q}_j \le 4 \operatorname{diam} \mathcal{Q}_j$. Then,

$$\operatorname{card} A_i^k = \sum_{j \in A_i^k} 1 \le \sum_{j \in A_i^k} \frac{\mathcal{L}(\mathcal{Q}_j)}{\mathcal{L}(\mathcal{Q}_i)} \le 40^n$$

and this completes the proof.

$\square$

Theorem 2.3.4 (Whitney Covering Theorem for (quasi-)metric spaces). *Let (E, d) be a (quasi-)metric space, and let $O \subsetneq E$ be open. Then there exists a set of balls $\mathcal{E} = \{B_\alpha\}_{\alpha \in I}$ and a constant $c_1 < \infty$ independent of O such that*

(i) The balls in $\mathcal{E}$ are mutually disjoint,

(ii) $O = \bigcup_{\alpha \in I} c_1 B_\alpha$,

(iii) $4 c_1 B_\alpha \not\subset O$.

Moreover, if (E, d) is separable then I can be taken at most countable.

Proof. We give the proof in the metric case $(A = 1)$ and leave it as an exercise to adapt when $A > 1$. Let $\delta(x) = \operatorname{dist}(x, {}^c O)$. When $x \in O$, since O is open, $\delta(x) > 0$. Fix $0 < \varepsilon < 1/2$ to be chosen later. Define

$$\mathcal{B} = \{B(x, \varepsilon \delta(x)) : x \in O\}$$

and let $\mathscr{E} = \{B_\alpha\}_{\alpha \in I} \subset \mathscr{B}$ maximal with mutually disjoint balls. The existence of $\mathscr{E}$ is guaranteed by Zorn's Lemma. Now we set $r_\alpha = \varepsilon\delta(x_\alpha)$ and $c_1 = 1/2\varepsilon$. Then,

$$4c_1 B_\alpha = B(x_\alpha, 4c_1 r_\alpha) = B(x_\alpha, 4 \cdot (1/2\varepsilon)\varepsilon\delta(x_\alpha)) = B(x_\alpha, 2\delta(x_\alpha)) \not\subset O.$$

This proves (i) and (iii).

Now, suppose there exists $x \in O \setminus \bigcup_{\alpha \in I} c_1 B_\alpha$. By maximality, there exists a $\beta \in I$ such that $\varnothing \neq B(x, r_x) \cap B(x_\beta, r_{x_\beta})$. In particular, this implies that

$$d(x, x_\beta) \le \varepsilon(\delta(x) + \delta(x_\beta)) \quad \text{and} \quad d(x, x_\beta) \ge \frac{1}{2}\delta(x_\beta)$$

and so

$$\delta(x_\beta) \le \frac{\varepsilon}{\frac{1}{2} - \varepsilon}\delta(x).$$

Now, trivially, $B(x_\beta, 2\delta(x_\beta)) \subset B(x, 2\delta(x_\beta) + d(x, x_\beta))$ and by the inequalities above,

$$2\delta(x_\beta) + d(x, x_\beta) \le \left[\frac{2\varepsilon}{\frac{1}{2} - \varepsilon} + \varepsilon + \frac{\varepsilon^2}{\frac{1}{2} - \varepsilon} \right] \delta(x).$$

Let $\varphi(\varepsilon)$ denote the quantity within the square brackets and by putting this together, we find that $B(x_\beta, 2\delta(x_\beta)) \subset B(x, \varphi(\varepsilon)\delta(x))$. Now, note that $\varphi(\varepsilon) \to 0$, so we can choose $0 < \varepsilon < 1/2$ such that $\varphi(\varepsilon) < 1$. For such a choice of ε, we find $B(x_\beta, 2\delta(x_\beta)) \subset B(x, \varphi(\varepsilon)\delta(x)) \subset O$. But this is a contradiction since $B(x_\beta, 2\delta(x_\beta)) \not\subset O$.

In the case where (E, d) is separable, there is a dense countable set D. Each B_α must contain at least one point of D: pick one called d_α. As the balls being disjoints, if $\alpha \neq \beta$ then $d_\alpha \neq d_\beta$. This created an injective map from $\mathscr{E}$ to D. Hence $\mathscr{E}$ is at most countable. $\qquad\square$

Proposition 2.3.5. *Assume that $(E, d) = (\mathbb{R}^n, d_E)$, where $d_E(x, y) = |x - y|$, then $\{c_1 B_\alpha\}_{\alpha \in I}$ possesses the bounded covering property.*

Proof. Fix $\alpha \in I$ and let $A_\alpha = \{\beta \in I : c_1 B_\alpha \cap c_1 B_\beta \neq \varnothing\}$. Now, take $\beta \in A_\alpha$ and fix $z \in c_1 B_\alpha \cap c_1 B_\beta$. Then,

$$d(z, x_\beta) \le \mathrm{rad}\, c_1 B_\beta = c_1 \, \mathrm{rad}\, B_\beta = \frac{1}{2\varepsilon}\varepsilon\delta(x_\beta) = \frac{1}{2}\delta(x_\beta).$$

By the triangle inequality,

$$\delta(x_\beta) = \mathrm{dist}(x_\beta, {}^cO) \le \mathrm{dist}(z, {}^cO) + d(x_\beta, z) \le \mathrm{dist}(z, {}^cO) + \frac{1}{2}\delta(x_\beta)$$

and we conclude

$$\frac{1}{2}\delta(x_\beta) \le \mathrm{dist}(z, {}^cO).$$

Furthermore,

$$\mathrm{dist}(z, {}^cO) \le \mathrm{dist}(x_\alpha, {}^cO) + d(z, x_\alpha) \le \delta(x_\alpha) + \frac{1}{2}\delta(x_\alpha) = \frac{3}{2}\delta(x_\alpha).$$

Combining these two estimates, and by symmetry we conclude that

$$\frac{1}{3} \leq \frac{\delta(x_\beta)}{\delta(x_\alpha)} \leq 3.$$

Let $\mathscr{E} = \{B(x_\beta, \frac{\varepsilon}{3}\delta(x_\alpha)) : \beta \in A_\alpha\}$. These balls are mutually disjoint by the previous inequality. Now,

$$d(x_\alpha, x_\beta) \leq d(x_\alpha, z) + d(x_\beta, z) \leq \frac{1}{2}\delta(x_\alpha) + \frac{1}{2}\delta(x_\beta) \leq \frac{1}{2}\delta(x_\alpha) + \frac{3}{2}\delta(x_\alpha) \leq 2\delta(x_\alpha).$$

Now, set

$$C = \frac{\frac{\varepsilon}{3} + 2}{\frac{\varepsilon}{3}},$$

and combining this with the estimate above, $B(x_\beta, \frac{\varepsilon}{3}\delta(x_\alpha)) \subset B(x_\alpha, C\frac{\varepsilon}{3}\delta(x_\alpha))$. So again, the volumes of the balls can be compared since with constant C since ε is fixed, and using a volume argument as in the proof of Besicovitch (Theorem 2.1.5) we attain a bound on card A_α depending only on dimension. $\qquad\square$

Remark 2.3.6. *We note that the preceding proposition can be proved for a quasi-metric space having the following structural property: There exists a constant $0 < C < \infty$ such that for all $R > 0$, the number of mutually disjoint balls of radius R contained in a ball of radius $2R$ is bounded by C. Such (quasi-)metric spaces are said to have geometric doubling property.*

Exercise 2.3.7. *Show that Vitali's set I_0 is at most countable on a quasi-metric space E with the geometric doubling property.*

Chapter 3

Maximal functions

3.1 Centred Maximal function on $\mathbb{R}^n$

We begin with the introduction of the classical Hardy-Littlewood maximal function.

Definition 3.1.1 (Hardy-Littlewood Maximal function). *Let μ be a reference measure, Borel, positive, and locally finite. Let ν be a second, positive, Borel measure. Define:*

$$\mathcal{M}_\mu(\nu)(x) = \sup_{r>0} \frac{1}{\mu(B(x,r))} \int_{B(x,r)} d\nu = \sup_{r>0} \frac{\nu(B(x,r))}{\mu(B(x,r))}$$

for each $x \in \mathbb{R}^n$. If $f \in \mathrm{L}^1_{\mathrm{loc}}(d\mu)$, then set $d\nu(x) = |f(x)|d\mu(x)$ and define

$$\mathcal{M}_\mu(f)(x) = \mathcal{M}_\mu(\nu)(x) = \sup_{r>0} \frac{1}{\mu(B(x,r))} \int_{B(x,r)} |f|\, d\mu.$$

Remark 3.1.2. *By convention, we take $\frac{0}{0} = 0$.*

The first and fundamental question is to ask the size of $\mathcal{M}_\mu(\nu)$ in terms of ν.

Theorem 3.1.3 (Maximal Theorem). *Suppose that ν is a finite measure. Then there exists a constant depending only on the dimension $C = C(n) > 0$ such that for all $\lambda > 0$,*

$$\mu\left\{x \in \mathbb{R}^n : \mathcal{M}_\mu(\nu)(x) > \lambda\right\} \leq \frac{C}{\lambda}\nu(\mathbb{R}^n).$$

Remark 3.1.4. *(i) The function $\mathcal{M}_\mu(\nu)$ is a Borel function.*

(ii) With some regularity assumptions on μ, $\mathcal{M}_\mu(\nu)$ becomes lower semi-continuous. That is, for all $\lambda > 0$, the set $\{x \in \mathbb{R}^n : \mathcal{M}_\mu(\nu)(x) > \lambda\}$ is open.

For example, suppose the map $(x,r) \mapsto \mu(B(x,r))$ is continuous (or equivalently $\mu(S^{n-1}(x,r)) = 0$ (Exercise)). Then, $\mathcal{M}_\mu(\nu)$ is lower semi-continuous (Exercise).

Proof of Maximal Theorem. Let $O_\lambda = \{x \in \mathbb{R}^n : \mathcal{M}_\mu(\nu)(x) > \lambda\}$. So, for all $x \in O_\lambda$, there exists $B_x = B(x, r_x)$ such that $\nu(B_x) > \lambda\mu(B_x)$. Fix $R > 0$, and apply Besicovitch to the set $O_\lambda \cap B(0, R)$. So, there is an $E \subset O_\lambda \cap B(0, R)$ at most countable such that

$$O_\lambda \cap B(0, R) \subset \bigcup_{x \in E} B_x \quad \text{and} \quad \sum_{x \in E} \chi_{B_x} \leq C(n).$$

Then,

$$\mu(O_\lambda \cap B(0, R)) \leq \sum_{x \in E} \mu(B_x) \leq \sum_{x \in E} \frac{1}{\lambda} \nu(B_x) = \sum_{x \in E} \frac{1}{\lambda} \int_{\mathbb{R}^n} \chi_{B_x} \, d\nu$$

$$= \frac{1}{\lambda} \int_{\mathbb{R}^n} \sum_{x \in E} \chi_{B_x} \, d\nu \leq \frac{C(n)}{\lambda} \nu(\mathbb{R}^n).$$

The sets $O_\lambda \cap B(0, R)$ are increasing as $R \to \infty$, so

$$\mu(O_\lambda) = \lim_{R \to \infty} \mu(O_\lambda \cap B(0, R)) \leq \frac{C(n)}{\lambda} \nu(\mathbb{R}^n)$$

which completes the proof. $\qquad\square$

Corollary 3.1.5. *For all $f \in L^1(d\mu)$ and for all $\lambda > 0$,*

$$\mu\{x \in \mathbb{R}^n : \mathcal{M}_\mu(f)(x) > \lambda\} \leq \frac{C(n)}{\lambda} \int_{\mathbb{R}^n} |f| \, d\mu.$$

Theorem 3.1.6 (Lebesgue Differentiation Theorem)**.** *Let $f \in L^1_{\mathrm{loc}}(d\mu)$. Then, there exists an $L_f \subset \mathbb{R}^n$ such that $\mu({}^c L_f) = 0$ and*

$$\lim_{r \to 0} \frac{1}{\mu(B(x, r))} \int_{B(x,r)} |f(x) - f(y)| \, d\mu(y) = 0$$

for all $x \in L_f$. In particular,

$$f(x) = \lim_{r \to 0} \frac{1}{\mu(B(x,r))} \int_{B(x,r)} f(y) d\mu(y)$$

for all $x \in L_f$.

Remark 3.1.7.   1. *This is a* local *statement, and so we can replace f by $f\chi_K$ where K is any compact set. Consequently we can assume that $f \in L^1(d\mu)$.*

2. *If f is continuous, then we can take $L_f = \varnothing$. That is, the Theorem holds everywhere.*

Proof. Define:

$$\omega_f(x) = \limsup_{r \to 0} \frac{1}{\mu(B(x, r))} \int_{B(x,r)} |f(x) - f(y)| \, d\mu(y)$$

and we remark that the measurability of ω_f is the same as the measurability of $\mathcal{M}_\mu(f)$. Note that ω_f is subadditive, that is, $\omega_{f+g} \leq \omega_f + \omega_g$. Also, $\omega_f \leq |f| + \mathcal{M}_\mu(f)$.

Let $\varepsilon > 0$, and since $C^0_c(\mathbb{R}^n)$ is dense in $L^1(d\mu)$ (since μ is locally finite), there exists a $g \in C^0_c(\mathbb{R}^n)$ such that

$$\int_{\mathbb{R}^n} |f - g| \, d\mu < \varepsilon.$$

Now, observe that from the previous remark (ii), $\omega_g = 0$ and so it follows that

$$\omega_f \le \omega_{f-g} + \omega_y \le |f - g| + \mathcal{M}_\mu(f - y).$$

Fix $\lambda > 0$ and note that

$$\{x \in \mathbb{R}^n : \omega_f(x) > \lambda\} \subset \left\{x \in \mathbb{R}^n : |(f - g)|(x) > \frac{\lambda}{2}\right\} \cup \left\{x \in \mathbb{R}^n : \mathcal{M}_\mu(f - g)(x) > \frac{\lambda}{2}\right\}.$$

From this, it follows that

$$\mu\left\{x \in \mathbb{R}^n : \omega_f(x) > \lambda\right\} \le \mu\left\{x \in \mathbb{R}^n : |(f - g)|(x) > \frac{\lambda}{2}\right\} +$$

$$\mu\left\{x \in \mathbb{R}^n : \mathcal{M}_\mu(f - g)(x) > \frac{\lambda}{2}\right\}$$

$$\le \frac{1}{\frac{\lambda}{2}} \int_{\mathbb{R}^n} |f - g| \; d\mu + \frac{C(n)}{\frac{\lambda}{2}} \int_{\mathbb{R}^n} |f - g| \; d\mu$$

$$\le \frac{1 + C(n)}{\frac{\lambda}{2}} \varepsilon.$$

Now letting $\varepsilon \to 0$, we find that $\mu\{x \in \mathbb{R}^n : \omega_f(x) > \lambda\} = 0$ and $\mu\{x \in \mathbb{R}^n : \omega_f(x) > 0\} = \lim_{n \to \infty} \mu\{x \in \mathbb{R}^n : \omega_f(x) > \frac{1}{n}\} = 0$. To complete the proof, we set $L_f = \{x \in \mathbb{R}^n : \omega_f(x) > 0\}$. $\square$

Remark 3.1.8 (On the measurability of ω_f). *Let*

$$\omega_{f;\frac{1}{k}}(x) = \limsup_{r \to 0, \; r \le \frac{1}{k}} \frac{1}{\mu(B(x,r))} \int_{B(x,r)} |f(x) - f(y)| \; d\mu(y)$$

and

$$\mathcal{M}_\mu^{\frac{1}{k}}(f)(x) = \sup_{r > 0, \; r \le \frac{1}{k}} \frac{1}{\mu(B(x,r))} \int_{B(x,r)} |f| \; d\mu.$$

Then, note that as $k \nearrow \infty$, $\omega_{f;\frac{1}{k}} \searrow \omega_f$. Now, for all $\rho \in \mathbb{Q} + \imath\mathbb{Q}$,

$$|f(x) - f(y)| \le |f(x) - \rho| + |\rho - f(y)|$$

and so it follows that

$$\omega_{f;\frac{1}{k}}(x) \le |f(x) - \rho| + \mathcal{M}_\mu^{\frac{1}{k}}(\rho - f)(x)$$

and also that

$$\omega_{f,\frac{1}{k}}(x) \le \inf_{\rho \in \mathbb{Q} + \imath\mathbb{Q}} \left\{|f(x) - \rho| + \mathcal{M}_\mu^{\frac{1}{k}}(\rho - f)(x)\right\}.$$

By the density of $\mathbb{Q} + \imath\mathbb{Q}$, equality holds for $x \in F$ where $F = \{x \in \mathbb{R}^n : f(x) \in \mathbb{C}\}$. Since $f \in \mathrm{L}^1(d\mu)$, we have $\mu({}^cF) = 0$.

Then,

$$\omega_f \text{ measurable} \iff \omega_{f;\frac{1}{k}} \text{ measurable} \quad \forall k$$

$$\iff \mathcal{M}_\mu^{\frac{1}{k}}(\rho - f) \text{ measurable} \quad \forall k \; \forall \rho$$

$$\iff \mathcal{M}_\mu(f) \text{ measurable}.$$

Exercise 3.1.9. *Prove that if $\mu(S^{n-1}(x,r)) = 0$ for all $x \in \mathbb{R}^n, r > 0$, then ω_f is Borel measurable.*

Remark 3.1.10. *The use of Besicovitch means that this is specific to $\mathbb{R}^n$.*

3.2 Maximal functions for doubling measures

Definition 3.2.1 (Doubling measure). *Let μ be a positive, locally finite, Borel measure. We say it is* doubling *if there exists a $C > 0$ such that for all $x \in \mathbb{R}^n$ and all $r > 0$,*

$$\mu(B(x, 2r)) \leq C\mu(B(x, r)).$$

Definition 3.2.2 (Uncentred maximal function). *Let μ be positive, locally finite, and Borel and ν positive and Borel. Define*

$$\mathcal{M}'_\mu(\nu)(x) = \sup_{B \ni x} \frac{\nu(B)}{\mu(B)}$$

for all $x \in \mathbb{R}^n$. As before, when $f \in \mathrm{L}^1_{\mathrm{loc}}(d\mu)$, $\mathcal{M}'_\mu(f) = \mathcal{M}'_\mu(\nu)$ where $d\nu = |f| \, d\mu$. That is,

$$\mathcal{M}'_\mu(f)(x) = \sup_{B \ni x} \frac{1}{\mu(B)} \int_B |f| \, d\mu.$$

Cheap Trick 3.2.3. *There exists a constant $C = C(n) > 0$ such that*

$$\mathcal{M}_\mu(\nu) \leq \mathcal{M}'_\mu(\nu) \leq C(n)\mathcal{M}_\mu(\nu).$$

Lemma 3.2.4. $\mathcal{M}'_\mu(\nu)$ *is a lower semi-continuous (and hence Borel) function.*

Proof. Fix $\lambda > 0$ and fix $x \in \{x \in \mathbb{R}^n : \mathcal{M}'_\mu(\nu)(x) > \lambda\}$. So, there exists a ball B with $x \in B$ such that

$$\frac{\nu(B)}{\mu(B)} > \lambda.$$

But for any $y \in B$, we have

$$\mathcal{M}'_\mu(\nu)(y) > \frac{\nu(B)}{\mu(B)} > \lambda$$

and so $B \subset \{x \in \mathbb{R}^n : \mathcal{M}'_\mu(\nu)(x) > \lambda\}$. This exactly means that $\{x \in \mathbb{R}^n : \mathcal{M}'_\mu(\nu)(x) > \lambda\}$ is open. $\qquad\square$

Theorem 3.2.5 (Maximal theorem for doubling measures). *Let μ be a doubling measure. Then, there exists an constant $C(n, \mu) > 0$ depending on dimension and the constant C in the doubling condition for μ such that for all $f \in \mathrm{L}^1(d\mu)$ and all $\lambda > 0$,*

$$\mu\{x \in \mathbb{R}^n : \mathcal{M}'_\mu(f)(x) > \lambda\} \leq \frac{C}{\lambda} \int_{\mathbb{R}^n} |f| \, d\mu.$$

Cheap proof. We use Cheap Trick 3.2.3 coupled with the Maximal theorem for $\mathcal{M}_\mu(f)$. $\qquad\square$

The preceding proof has the disadvantage that it is inherently tied up with $\mathbb{R}^n$. The following is a better proof.

Better proof. Define: $O'_\lambda = \{x \in \mathbb{R}^n : \mathcal{M}'_\mu(f)(x) > \lambda\}$,

$$\mathcal{M}'^m_\mu(\nu)(x) = \sup_{B \ni x, \text{ rad } B \leq m} \frac{\nu(B)}{\mu(B)},$$

and $O'^m_\lambda = \{x \in \mathbb{R}^n : \mathcal{M}'^m_\mu(f)(x) > \lambda\}$. Then, $O'_\lambda = \bigcup_{m=1}^\infty O'^m_\lambda$.

Fix m. Then, for all $x \in O'^m_\lambda$, there exists a ball B_x with $x \in B_x$ with rad $B_x \leq m$ such that

$$\frac{1}{\mu(B_x)} \int_{B_x} |f| \ d\mu > \lambda.$$

By a repetition of the argument in Lemma 3.2.4, $B \subset O'^m_\lambda$ making $\mathcal{M}'_\mu(f)$ Borel.

Let $\mathscr{B} = \{B_x\}_{x \in O'^m_\lambda}$, and apply the Vitali Covering Lemma 2.1.1. So, there exists a countable subset of centres $\mathcal{C} \subset O'^m_\lambda$ and mutually disjoint $\{B_{x_j}\}_{x_j \in \mathcal{C}} \subset \mathscr{B}$ satisfying

$$O'^m_\lambda \subset \bigcup_{x_j \in C} 5B_{x_j}.$$

Therefore,

$$\mu(O'^m_\lambda) \leq \sum_{x_j \in \mathcal{C}} \mu(5B_{x_j}) \leq C^3 \sum_{x_j \in \mathcal{C}} \mu(B_{x_j})$$

$$\leq \frac{C^3}{\lambda} \sum_{x_j \in C} \int_{B_{x_j}} |f| \ d\mu = \frac{C^3}{\lambda} \int_{\bigcup_{x_j \in C} B_{x_j}} |f| \ d\mu \leq \frac{C^3}{\lambda} \int_{O'^m_\lambda} |f| \ d\mu \leq \frac{C^3}{\lambda} \int_{O'_\lambda} |f| \ d\mu$$

Now, O'^m_λ is an increasing sequence of sets and so we obtain the desired conclusion letting $m \to \infty$. $\qquad\square$

Remark 3.2.6. *This proof is not only better because it frees itself from Besicovitch to the more general Vitali, but we get a better estimate since the integral is over O'_λ rather than $\mathbb{R}^n$.*

Corollary 3.2.7 (Lebesgue Differentiation Theorem for doubling measures). *We have*

$$f(x) = \lim_{B \ni x, \text{ rad } B \to 0} \frac{1}{\mu(B)} \int_B f(y) \ d\mu(y)$$

for μ-almost everywhere $x \in \mathbb{R}^n$.

3.3   The Dyadic Maximal function

Let μ be a positive, locally finite Borel measure.

Definition 3.3.1 (Dyadic Maximal function). *Let $\mathcal{Q}_0 \in \mathscr{Q}$ and let $\mathscr{D}(\mathcal{Q}_0)$ be the collection of dyadic subcubes of $\mathcal{Q}_0$. Define:*

$$\mathcal{M}^{\mathscr{Q}}_\mu(f)(x) = \sup_{\mathcal{Q} \in \mathscr{D}(\mathcal{Q}_0), \ \mathcal{Q} \ni x} \frac{1}{\mu(\mathcal{Q})} \int_\mathcal{Q} |f| \ d\mu$$

for $x \in \mathcal{Q}_0$ and $f \in \mathrm{L}^1_{\mathrm{loc}}(\mathcal{Q}_0; d\mu)$.

Theorem 3.3.2 (Maximal theorem for the dyadic maximal function). *Suppose* $f \in L^1(\mathcal{Q}_0; d\mu)$ *and* $\lambda > 0$. *Then,*

$$\mu\left\{x \in \mathcal{Q}_0 : \mathcal{M}_\mu^{\mathscr{D}}(f)(x) > \lambda\right\} \leq \frac{1}{\lambda} \int_{\mathcal{Q}_0} |f| \ d\mu.$$

Remark 3.3.3. *Notice here that the bounding constant here is* 1. *It is independent of dimension and* μ.

We give two proofs.

Proof 1. Let $\Omega_\lambda = \left\{x \in \mathcal{Q}_0 : \mathcal{M}_\mu^{\mathscr{D}}(f)(x) > \lambda\right\}$. If $\Omega_\lambda \neq \varnothing$ and $x \in \Omega_\lambda$, there exists a $\mathcal{Q} \in \mathscr{D}(\mathcal{Q}_0)$ such that $x \in \mathcal{Q}$ and

$$\frac{1}{\mu(\mathcal{Q})} \int_{\mathcal{Q}} |f| \ d\mu > \lambda. \tag{$\dagger$}$$

Let $\mathscr{C}$ be the collection of all such cubes. Since $\mathscr{D}$ is countable, $\mathscr{C}$ is also countable. Then, for every $\mathcal{Q} \in \mathscr{C}$, we have that $\mathcal{Q} \subset \Omega_\lambda$ so therefore, $\Omega_\lambda = \bigcup_{\mathcal{Q} \in \mathscr{C}} \mathcal{Q}$. Let $\mathscr{M} = \{\mathcal{Q}_i\}_{i \in I} \subset \mathscr{C}$ be a maximal subcollection. Then, $\mathscr{M}$ is a partition of $\mathcal{Q}_0$. So,

$$\mu(\Omega_\lambda) = \sum_{i \in I} \mu(\mathcal{Q}_i) \leq \frac{1}{\lambda} \sum_{i \in I} \int_{\mathcal{Q}_i} |f| \ d\mu \leq \frac{1}{\lambda} \int_{\Omega_\lambda} |f| \ d\mu.$$

$\square$

Remark 3.3.4.   (i) *The uniqueness of* $\mathscr{M}$ *is a consequence of the disjointness of the dyadic cubes at each generation.*

   (ii) *The proof gives us an even sharper inequality since we are only integrating on the set* Ω_λ *rather than* $\mathcal{Q}_0$.

Proof 2. If

$$\frac{1}{\mu(\mathcal{Q}_0)} \int_{\mathcal{Q}_0} |f| \ d\mu > \lambda,$$

then $\Omega_\lambda = \{\mathcal{Q}_0\}$ and there's nothing to do. So assume the converse. We construct a mutually disjoint subset $\mathscr{F}$ by the following procedure. Consider a dyadic child of $\mathcal{Q}_0$, say, $\mathcal{Q}$. If $\mathcal{Q}$ satisfies ($\dagger$), we stop and put $\mathcal{Q}$ in $\mathscr{F}$. Otherwise, we apply this procedure to $\mathcal{Q}$ in place of $\mathcal{Q}_0$. That is, we consider whether a given dyadic child of $\mathcal{Q}$ to satisfies ($\dagger$). The collection $\mathscr{F}$ is called the *stopping cubes* for the property ($\dagger$). [1]

We claim that:

$$\Omega_\lambda = \bigcup_{\mathcal{Q} \in \mathscr{F}} \mathcal{Q}.$$

Clearly, $\mathcal{Q} \in \mathscr{F}$ implies that $\mathcal{Q} \subset \Omega_\lambda$ (Proof 1). Suppose that there exists $x \in \Omega_\lambda$ but $x \notin \bigcup_{\mathcal{Q} \in \mathscr{F}} \mathcal{Q}$. But we know that $\mathcal{M}_\mu^{\mathscr{D}}(f)(x) > \lambda$ so there exists a dyadic cube $\mathcal{Q}' \in \mathscr{D}(\mathcal{Q}_0)$ such that $x \in \mathcal{Q}'$ satisfying the property ($\dagger$). The stopping time did not stop before $\mathcal{Q}'$ by

---

[1] This is a stopping time argument, a typical technique in probability.

hypothesis $x \notin \bigcup_{Q \in \mathscr{F}} Q$, but Q' satisfies (†). By the construction of $\mathscr{F}$, we have $Q' \in \mathscr{F}$ which is a contradiction.

The estimate then follows by the same calculation as in Proof 1. $\qquad \square$

Remark 3.3.5. *The collection $\mathscr{F}$ is the maximal collection $\mathscr{M}$ from Proof 1.*

Definition 3.3.6 (Dyadic maximal function on $\mathbb{R}^n$)**.** *Define:*

$$\mathcal{M}_\mu^{\mathscr{D}}(f)(x) = \sup_{Q \in \mathscr{D}, \ Q \ni x} \frac{1}{\mu(Q)} \int_Q |f| \ d\mu$$

for $x \in \mathbb{R}^n$ and $f \in \mathrm{L}^1_{\mathrm{loc}}(d\mu)$.

Corollary 3.3.7. *Suppose $f \in \mathrm{L}^1(\mathbb{R}^n; d\mu)$ and $\lambda > 0$. Then,*

$$\mu\left\{x \in \mathbb{R}^n : \mathcal{M}_\mu^{\mathscr{D}}(f)(x) > \lambda\right\} \leq \frac{1}{\lambda} \int_{\mathbb{R}^n} |f| \ d\mu.$$

Proof. Since we are considering dyadic cubes, we begin by splitting $\mathbb{R}^n$ into quadrants, and let $g = f \chi_{(\mathbb{R}_+)^n}$. We note that it suffices to prove the statement for one quadrant since the argument is unchanged for the others.

For each $k \in \mathbb{N}$, let $Q^k = [0, 2^k)^n$. Then, note that:

$$\mathcal{M}_\mu^{\mathscr{D}}(f)(x) = \sup_{k \in \mathbb{N}} g_k(x)$$

where

$$g_k(x) = \sup_{Q \in \mathscr{D}(Q^k), \ Q \subset Q^k, \ Q \ni x} \frac{1}{\mu(Q)} \int_Q |f| \ d\mu.$$

Now, let $\Omega_\lambda^k = \left\{x \in Q^k : g_k(x) > \lambda\right\}$. We compute:

$$\mu\left\{x \in (\mathbb{R}_+)^n : g_k(x) > \lambda\right\} = \mu(\Omega_\lambda^k) \leq \frac{1}{\lambda} \int_{Q^k} |f| \ d\mu \leq \frac{1}{\lambda} \int_{(\mathbb{R}_+)^n} |f| \ d\mu.$$

The desired conclusion is achieved by letting $k \to \infty$ and summing over all quadrants. $\qquad \square$

Corollary 3.3.8. *For $f \in \mathrm{L}^1_{\mathrm{loc}}(d\mu)$, we have*

$$f(x) = \lim_{Q \in \mathscr{D}, \ \mathscr{L}(Q) \to 0, Q \ni x} \frac{1}{\mu(Q)} \int_Q f(y) \ d\mu(y)$$

for μ-almost everywhere $x \in \mathbb{R}^n$.

Proof. Exercise. $\qquad \square$

Remark 3.3.9. *For a fixed x, the set $\{Q \in \mathscr{D} : \mathscr{L}(Q) \to 0, Q \ni x\}$ is really a sequence. Consequently, the limit is really the limit of a sequence.*

We have the following important Application.

Proposition 3.3.10. *For $f \in \mathrm{L}^1_{\mathrm{loc}}(d\mu)$,*

$$|f| \leq \begin{cases} \mathcal{M}^{\mathscr{D}}_\mu(f) \\ \mathcal{M}_\mu(f) \\ \mathcal{M}'_\mu(f) \end{cases}$$

for μ-almost everywhere $x \in \mathbb{R}^n$.

Remark 3.3.11. *The centred maximal function uses Besicovitch and is confined to $\mathbb{R}^n$. The others do not and can be generalised to spaces with appropriate structure. For instance, in the case of the Dyadic maximal function, we must be able to at least perform a dyadic decomposition of the space.*

A consequence of the Lebesgue differentiation is:

$$\mathcal{M}_\mu(f) \leq \mathcal{M}'_\mu(f).$$

Exercise 3.3.12. *Try to compare $\mathcal{M}^{\mathscr{D}}_\mu$ with $\mathcal{M}_\mu$ (and $\mathcal{M}'_\mu$). For simplicity, take $\mu = \mathscr{L}$ (but note that this has nothing to do with the measure but rather the geometry).*

3.4 Maximal Function on L^p spaces

In this section, we let $\mathcal{M}$ denote either $\mathcal{M}_\mu, \mathcal{M}'_\mu$ or $\mathcal{M}^{\mathscr{D}}_\mu$. We firstly note that for every $f \in \mathrm{L}^\infty(d\mu)$,

$$\mathcal{M}f(x) \leq \|f\|_\infty$$

for all $x \in \mathbb{R}^n$. So, $\mathcal{M}f$ is a bounded operator on $\mathrm{L}^\infty(d\mu)$. We investigate the boundedness on $\mathrm{L}^p(d\mu)$ for $p < \infty$.

Lemma 3.4.1 (Cavalieri's principle). *Let $0 < p < \infty$ and let g be a positive, measurable function. Then,*

$$\int_{\mathbb{R}^n} g^p \, d\mu = p \int_0^\infty \mu \{x \in \mathbb{R}^n : g(x) > \lambda\} \lambda^{p-1} \, d\lambda.$$

Proof. Assume that μ finite (this is required for the application of Fubini's theorem in the following computation). In the case μ is not finite, assuming that $\mu \{x \in \mathbb{R}^n : g(x) > \lambda\} < \infty$ for all $\lambda > 0$, we can restrict μ to a σ-algebra (depending on g) where it is the case. Then this reduces to the assumption μ is finite by approximating g via $g_{N,R} = g\chi_{\{x \in \mathbb{R}^n : g(x) \leq N\}}\chi_{B(0,R)}$ and then applying the Monotone Convergence Theorem.

We compute and apply Fubini:

$$p \int_0^\infty \mu \{x \in \mathbb{R}^n : g(x) > \lambda\}\lambda^{p-1} \, d\lambda$$

$$= p \int_0^\infty \int_{\{x \in \mathbb{R}^n : g(x) > \lambda\}} 1 \, d\mu(x)\lambda^{p-1} \, d\lambda$$

$$= \int_{\mathbb{R}^n} \int_0^{g(x)} p\lambda^{p-1} \, d\lambda \, d\mu(x)$$

$$= \int_{\mathbb{R}^n} g^p \, d\mu(x).$$

$\square$

Theorem 3.4.2 (Boundedness of the Maximal function on L^p). *Let $1 < p < \infty$. Then, there exists a constant $C = C(p, n) > 0$ such that whenever $f \in \mathrm{L}^p(d\mu)$ then $\mathcal{M}f \in \mathrm{L}^p(d\mu)$ and*

$$\|\mathcal{M}f\|_{\mathrm{L}^p(d\mu)} \le C\|f\|_{\mathrm{L}^p(d\mu)}.$$

For the case of $\mathcal{M} = \mathcal{M}'_\mu$, we assume that μ is doubling and C may also depend on the constants in the doubling condition.

Proof. Assume that $f \in \mathrm{L}^1(d\mu) \cap \mathrm{L}^p(d\mu)$. Let $\lambda > 0$ and define

$$f_\lambda(x) = \begin{cases} f(x) & |f(x)| > \frac{\lambda}{2} \\ 0 & |f(x)| \le \frac{\lambda}{2} \end{cases}$$

and

$$|f| \le |f_\lambda| + \frac{\lambda}{2}.$$

By the subadditivity of the supremum,

$$\mathcal{M}f \le \mathcal{M}f_\lambda + \frac{\lambda}{2}$$

and

$$\{x \in \mathbb{R}^n : \mathcal{M}f(x) > \lambda\} \subset \left\{x \in \mathbb{R}^n : \mathcal{M}f_\lambda(x) > \frac{\lambda}{2}\right\}.$$

Therefore,

$$\mu\{x \in \mathbb{R}^n : \mathcal{M}f(x) > \lambda\} \le \mu\left\{x \in \mathbb{R}^n : \mathcal{M}f_\lambda(x) > \frac{\lambda}{2}\right\} \le \frac{C}{\lambda} \int_{\mathbb{R}^n} |f_\lambda| \ d\mu.$$

Then,

$$p \int_0^\infty \mu\{x \in \mathbb{R}^n : \mathcal{M}f(x) > \lambda\} \lambda^{p-1} \ d\lambda \le p \int_0^\infty C \int_{\mathbb{R}^n} |f_\lambda| \lambda^{p-2} \ d\mu d\lambda$$

$$\le Cp \int_{\mathbb{R}^n} \int_0^{2f(x)} |f_\lambda| \lambda^{p-2} \ d\lambda d\mu$$

$$\le C2^{p-1} \frac{p}{p-1} \int_{\mathbb{R}^n} |f|^p \ d\mu$$

Now, for a general $f \in \mathrm{L}^p(d\mu)$ by the density of $\mathrm{L}^1(d\mu) \cap \mathrm{L}^p(d\mu)$ in $\mathrm{L}^p(d\mu)$, we can take a sequence (for example simple functions) $f_k \in \mathrm{L}^1(d\mu) \cap \mathrm{L}^p(d\mu)$ satisfying $|f_k| \nearrow |f|$. Then, $\mathcal{M}f_k \nearrow \mathcal{M}f$ and the proof is complete by invoking the Monotone Convergence Theorem. $\square$

Remark 3.4.3. *(i) The constant $C(n, p)$ satisfies:*

$$C(n, p) \le \left(C2^{p-1} \frac{p}{p-1}\right)^{\frac{1}{p}} \tag{†}$$

So, if we let $p \to \infty$, then we find that

$$\left(C2^{p-1} \frac{p}{p-1}\right)^{\frac{1}{p}} \to 1.$$

27

But as $p \to 1$, $C \sim \frac{p}{p-1}$ and blows up!

In fact it is true that if $0 \neq f \in \mathrm{L}^1(d\mathscr{L})$, then $\mathcal{M}f \notin \mathrm{L}^1(d\mathscr{L})$ (Exercise). For $p = 1$, the best result is the Maximal Theorem.

(ii) *We remark on the optimality of $C(n,p)$. Consider the inequality (†), and note that the C is the Maximal Theorem constant. This depends on n and the measure μ (unless we consider $\mathcal{M}_\mu^{\mathcal{Q}}$).*

Suppose $\mu = \mathscr{L}$. Then the best upper bound for C with respect to n is $n \log n$. Consider the operator norm of $\mathcal{M}$:

$$M_{p,n} = \sup_{f \in \mathrm{L}^p(d\mathscr{L}), \ \|f\|_{\mathrm{L}^p(d\mathscr{L})}=1} \|\mathcal{M}f\|_{\mathrm{L}^p(d\mathscr{L})}.$$

If $\mathcal{M} = \mathcal{M}_\mu$ (ie, the centred maximal function), then $M_{p,n}$ can be shown to be bounded in n for any fixed $p > 1$. This is important in stochastic analysis.

Now, suppose $\mathcal{M} = \mathcal{M}_\mu^{\mathcal{Q}}$. Then,

$$M_{p,n} = \frac{p}{p-1}.$$

To see this, first note that

$$\mathscr{L}\{x \in \mathbb{R}^n : \mathcal{M}f(x) > \lambda\} \leq \frac{1}{\lambda} \int_{\{x \in \mathbb{R}^n : \mathcal{M}f(x) > \lambda\}} |f| \ d\mathscr{L}$$

and

$$\begin{aligned}
\|\mathcal{M}f\|_{\mathrm{L}^p(d\mathscr{L})}^p &= \int_{\mathbb{R}^n} |\mathcal{M}f|^p \ d\mathscr{L} \\
&= p \int_0^\infty \mathscr{L}\{x \in \mathbb{R}^n : \mathcal{M}f(x) > \lambda\} \lambda^{p-1} \ d\lambda \\
&\leq p \int_0^\infty \int_{\{x \in \mathbb{R}^n : \mathcal{M}f(x) > \lambda\}} |f| \ d\mathscr{L} \lambda^{p-1} \ d\lambda \\
&= p \int_0^\infty \left(\int_0^{\mathcal{M}f} \lambda^{p-2} \ d\lambda \right) |f| \ d\mathscr{L} \\
&= \frac{p}{p-1} \int_{\mathbb{R}^n} (\mathcal{M}f)^{p-1} |f| \ d\mathscr{L} \\
&\leq \frac{p}{p-1} \|\mathcal{M}f\|_{\mathrm{L}^p(d\mathscr{L})}^{p-2} \|f\|_{\mathrm{L}^p(d\mathscr{L})}^p
\end{aligned}$$

which shows that

$$\|\mathcal{M}f\|_{\mathrm{L}^p(d\mathscr{L})} \leq \frac{p}{p-1} \|f\|_{\mathrm{L}^p(d\mathscr{L})}.$$

Optimality can be shown via martingale techniques. We will not prove this.

We give the following important application of Maximal function theory.

Theorem 3.4.4 (Hardy-Littlewood Sobolev inequality). *Let $0 < \lambda < n$, $1 \leq p < \frac{n}{n-\lambda}$ and q satisfying*

$$\frac{1}{p} + \frac{\lambda}{n} = 1 + \frac{1}{q}.$$

Then, with $\nu_\lambda = \frac{1}{|\cdot|^\lambda}$,

(i) When $p = 1$, there exists a constant $C(\lambda, n)$ such that for all $\alpha > 0$,

$$\mathscr{L} \{x \in \mathbb{R}^n : |\nu_\lambda * u(x)| > \alpha\} \leq \frac{C}{\alpha^{\frac{n}{\lambda}}} \|u\|_{L^1(d\mathscr{L})}^{\frac{n}{\lambda}},$$

(ii) When $p > 1$, there exists a constant $C(p, \lambda, n)$ such that

$$\|\nu_\lambda * u\|_{L^q(d\mathscr{L})} \leq C\|u\|_{L^p(d\mathscr{L})},$$

where

$$(\nu_\lambda * u)(x) = \int_{\mathbb{R}^n} \frac{1}{|x-y|^\lambda} u(y) \, d\mathscr{L}(y)$$

is the convolution of u with ν_λ.

Remark 3.4.5 (Motivation). *Such a ν_λ arises when trying to "integrate" functions in $\mathbb{R}^n$. Such "potentials" are one way to "anti-derive." Formally, in the case of $\lambda = n - 2$,*

$$\left(\frac{1}{|x|^{n-2}} * u\right)\widehat{}(\xi) = C(n) \frac{\hat{u}(\xi)}{|\xi|^2}$$

where $\widehat{}$ denotes the Fourier Transform. *See [Ste71a].*

Proof (Hedberg's inequality). We prove (i). Let $u \in C_c^\infty(\mathbb{R}^n)$, and set $w = \nu_\lambda * u$. We take $\lambda > 0$ to be chosen later. Then,

$$w(x) \leq \int_{|x-y|\geq\delta} \frac{1}{|x-y|^\lambda} |u(y)| \, d\mathscr{L}(y) + \sum_{k=0}^{\infty} \int_{\delta 2^{-k-1}\leq|x-y|\leq\delta 2^{-k}} \frac{1}{|x-y|^\lambda} |u(y)| \, d\mathscr{L}(y).$$

First, whenever $|x - y| \geq \delta$,

$$\int_{|x-y|\geq\delta} \frac{1}{|x-y|^\lambda} |u(y)| \, d\mathscr{L}(y) \leq \frac{1}{\delta^\lambda} \|u\|_{L^1(d\mathscr{L})}.$$

So, fix $k \geq 0$. Then,

$$\int_{\delta 2^{-k-1}\leq|x-y|\leq\delta 2^{-k}} \frac{1}{|x-y|^\lambda} |u(y)| \, d\mathscr{L}(y)$$

$$\leq (\delta 2^{-k-1})^{-\lambda} \frac{\mathscr{L}(B(x, \delta 2^{-k}))}{\mathscr{L}(B(x, \delta 2^{-k}))} \int_{|x-y|\leq\delta 2^{-k}} |u(y)| \, d\mathscr{L}(y)$$

$$\leq b_n 2^\lambda (\delta 2^{-k})^{n-\lambda} \mathcal{M}_{\mathscr{L}}(u)(x)$$

where $b_n = \mathscr{L}(B(0, 1))$, the volume of the n ball. It follows then that

$$\sum_{k=0}^{\infty} \int_{\delta 2^{-k-1}\leq|x-y|\leq\delta 2^{-k}} \frac{1}{|x-y|^\lambda} |u(y)| \, d\mathscr{L}(y)$$

$$\leq \sum_{k=0}^{\infty} b_n 2^\lambda (\delta 2^{-k})^{n-\lambda} \mathcal{M}_{\mathscr{L}}(u)(x)$$

$$\leq \delta^{-\lambda} \delta^n \frac{b_n 2^\lambda}{1 - 2^{-(n-\lambda)}} \mathcal{M}_{\mathscr{L}}(u)(x).$$

Now, let

$$A = A(n, \lambda) = \frac{b_n 2^\lambda}{1 - 2^{-(n-\lambda)}}$$

and solve for δ such that $\delta^n A(n, \lambda) \mathcal{M}_\mathscr{L}(u)(x) = \|u\|_{L^1(d\mathscr{L})}$. Since if $u \not\equiv 0$ then $\mathcal{M}_\mathscr{L} u(x) \neq 0$ almost everywhere (which we leave as an exercise),

$$\delta = \left(\frac{\|u\|_{L^1(d\mathscr{L})}}{A(n, \lambda) \mathcal{M}_\mathscr{L}(u)(x)} \right)^{\frac{1}{n}}.$$

Then, for almost everywhere $x \in \mathbb{R}^n$,

$$
\begin{aligned}
w(x) &\leq \delta^{-\lambda} \|u\|_{L^1(d\mathscr{L})} + \delta^{-\lambda} \delta^n A(n, \lambda) \mathcal{M}_\mathscr{L} u(x) \\
&= \delta^{-\lambda} 2 \|u\|_{L^1(d\mathscr{L})} \\
&= 2 \left(\frac{\|u\|_{L^1(d\mathscr{L})}}{A(n, \lambda) \mathcal{M}_\mathscr{L}(u)(x)} \right)^{-\frac{\lambda}{n}} \|u\|_{L^1(d\mathscr{L})} \\
&= 2 \left(\frac{A(n, \lambda) \mathcal{M}_\mathscr{L}(u)(x)}{\|u\|_{L^1(d\mathscr{L})}} \right)^{\frac{1}{q}} \|u\|_{L^1(d\mathscr{L})} \\
&= A(n, \lambda)^{\frac{1}{q}} \mathcal{M}_\mathscr{L}(u)(x)^{\frac{1}{q}} \|u\|_{L^1(d\mathscr{L})}^{1-\frac{1}{q}}.
\end{aligned}
$$

From this, it follows that,

$$
\begin{aligned}
\mathscr{L}\{x \in \mathbb{R}^n : w(x) > \alpha\} &\leq \mathscr{L}\left\{ x \in \mathbb{R}^n : \mathcal{M}_\mathscr{L}(u)(x) > A(n, \lambda)^{-1} \alpha^q \left(\|u\|_{L^1(d\mathscr{L})}^{1-\frac{1}{q}} \right)^{-q} \right\} \\
&\leq \frac{C(n) A(n, \lambda)}{\alpha^q} \left(\|u\|_{L^1(d\mathscr{L})}^{1-\frac{1}{q}} \right)^q \|u\|_{L^1(d\mathscr{L})} \\
&= \frac{C(n) A(n, \lambda)}{\alpha^q} \|u\|_{L^1(d\mathscr{L})}^q.
\end{aligned}
$$

The density of $C_c^\infty(\mathbb{R}^n)$ in $L^1(d\mathscr{L})$ completes the proof.

We leave (ii) as an exercise. $\qquad\square$

Chapter 4

Interpolation

4.1 Real interpolation

Suppose that (M, μ) and (N, ν) are measure spaces and the measures μ and ν are σ-finite measures.

Recall that when $0 < p < \infty$, $\mathrm{L}^p(M, \mu)$ (or $\mathrm{L}^p(M, d\mu)$) is the space of μ-measurable functions f for which $|f|^p$ is integrable. When $1 \leq p < \infty$, this space is a Banach space (modulo the almost everywhere equality) and the norm is what we expect:

$$\|f\|_p = \left(\int_M |f|^p \, d\mu \right)^{1/p}.$$

For $p = \infty$, $f \in \mathrm{L}^\infty(M, \mu)$ if and only if there exists a $\lambda > 0$ such that $\mu \{x \in M : |f(x)| > \lambda\} = 0$. This is a Banach space and the norm is then given by

$$\|f\|_\infty = \operatorname{esssup} |f| = \inf \{\lambda > 0 : \mu \{x \in M : |f(x)| > \lambda\} = 0\}.$$

We define a generalisation of these spaces called the *Weak* L^p *spaces*.

Definition 4.1.1 (Weak L^p space). *Let* $0 < p < \infty$. *Then, define* $\mathrm{L}^{p,\infty}(M, \mu)$ *to be the space of* μ-measurable functions f satisfying

$$\sup_{\lambda > 0} \lambda^p \mu \{x \in M : |f(x)| > \lambda\} < \infty.$$

For a function $f \in \mathrm{L}^{p,\infty}(M, \mu)$, *we define a quasi-norm:*

$$\|f\|_{p,\infty} = \left(\sup_{\lambda > 0} \lambda^p \mu \{x \in M : |f(x)| > \lambda\} \right)^{\frac{1}{p}}.$$

When $p = \infty$, *we let* $\mathrm{L}^{p,\infty}(M, \mu) = \mathrm{L}^\infty$.

Remark 4.1.2. *(i) The Weak* L^p *spaces are really a special case of* Lorentz spaces $\mathrm{L}^{p,q}$. *They truly generalise the* L^p *spaces for they are equal to* L^p *when* $q = p$. *See [Ste71b] for a detailed treatment.*

31

(ii) We emphasise that for $0 < p < \infty$, $\|f\|_{p,\infty}$ is not a norm, but it is a quasi-norm. That is,

$$\|f + g\|_{p,\infty} \leq 2(\|f\|_{p,\infty} + \|g\|_{p,\infty}).$$

This is a direct consequence of the following observation:

$$\{x \in M : |f + g| > \lambda\} \subset \left\{x \in M : |f| > \frac{\lambda}{2}\right\} \cup \left\{x \in M : |g| > \frac{\lambda}{2}\right\}.$$

(iii) When $1 < p < \infty$, there exists a norm on $L^{p,\infty}(M, \mu)$ which is metric equivalent to $\|\cdot\|_{p,\infty}$.

(iv) When $p = 1$, there is no such equivalent norm. In fact, $L^{1,\infty}(M, \mu)$ is not even locally convex. For instance when $u \in L^1(\mathbb{R}^n, d\mathscr{L})$ then $\mathcal{M}_{\mathscr{L}} u \in L^{1,\infty}(\mathbb{R}^n, d\mathscr{L})$ by the Maximal theorem.

Proposition 4.1.3. The Weak L^p spaces are complete with respect to the metric $d_p(f, g) = \|f - g\|_{p,\infty}$ for $0 < p < \infty$.

Proposition 4.1.4 (Tchebitchev-Markov inequality). If f is positive, μ-measurable and $0 < p < \infty$, then

$$\mu\{x \in M : f(x) > \lambda\} \leq \frac{1}{\lambda^p} \int_{\{x \in M : f(x) > \lambda\}} f^p \, d\mu(x) \leq \frac{1}{\lambda^p} \|f\|_p^p.$$

In particular, $L^p(M, \mu) \subset L^{p,\infty}(M, \mu)$.

Remark 4.1.5. The inclusion is strict in general. For instance,

$$\frac{1}{|x|^\lambda} \in L^{\frac{n}{\lambda},\infty}(\mathbb{R}^n, d\mathscr{L}) \setminus L^{\frac{n}{\lambda}}(\mathbb{R}^n, d\mathscr{L}),$$

when $0 < \lambda$.

Remark 4.1.6 (σ-finiteness of μ). The above definitions do not require σ-finiteness of μ. We will, however, require this in the interpolation theorem to follow.

Definition 4.1.7 (Sublinear operator). Let $\mathbb{K} = \mathbb{R}$ or $\mathbb{K} = \mathbb{C}$, and let $\mathscr{F}_M$ denote the space measurable functions $f : M \to \mathbb{K}$. Let $\mathcal{D}_M$ be a subspace of $\mathscr{F}_M$ and let $T : \mathcal{D}_M \to \mathscr{F}_N$. We say that T is sublinear if

$$|T(f_1 + f_2)(x)| \leq |Tf_1(x)| + |Tf_2(x)|$$

for ν-almost all $x \in N$.

Example 4.1.8. *(i)* $T : f \mapsto \mathcal{M}_{\mathscr{L}} f$ where $\mathcal{D} = L^1_{\text{loc}}(\mathbb{R}^n, d\mathscr{L})$.

(ii) Any linear T is sublinear.

Definition 4.1.9 (Weak/Strong type). Let $T : \mathcal{D}_M \to \mathscr{F}_N$ be sublinear and let $1 \leq p, q \leq \infty$. Then,

(i) T is of strong type (p, q) if $T : \mathcal{D}_M \cap L^p(M, \mu) \to L^q(N, \mu)$ is a bounded map. That is, there exists a $C > 0$ such that whenever $f \in \mathcal{D}_M \cap L^p(M, \mu)$ we have $Tf \in L^q(N, \mu)$ and

$$\|Tf\|_q \leq C\|f\|_p.$$

(ii) T *is of weak type* (p,q) *(for* $q < \infty$*) if* $T : \mathcal{D}_M \cap \mathrm{L}^p(M,\mu) \to \mathrm{L}^{q,\infty}(N,\mu)$ *is a bounded map. That is, there exists a* $C > 0$ *such that whenever* $f \in \mathcal{D}_M \cap \mathrm{L}^p(M,\mu)$ *we have* $Tf \subset \mathrm{L}^{q,\infty}(N,\mu)$ *and*

$$\|Tf\|_{q,\infty} \leq C\|f\|_p.$$

(iii) T *is of weak type* (p,∞) *if it is of strong type* (p,∞).

Remark 4.1.10. *Note that if* T *is of strong type* (p,q) *then it is of weak type* (p,q).

Example 4.1.11. *(i) The maximal operator* $T : f \to \mathcal{M}_\mu f$ *is of weak type* $(1,1)$ *and of strong type* (p,p) *for* $1 < p \leq \infty$.

(ii) The operator $u \mapsto \nu_\lambda * u$ *where* ν_λ *is the potential in Theorem 3.4.4 is of weak type* $(1, \frac{n}{\lambda})$ *if* $0 < \lambda < n$ *on* $(\mathbb{R}^n, \mathcal{L})$.

Theorem 4.1.12 (Marcinkiewicz Interpolation Theorem). *Given* (M,μ) *and* (N,ν) *let* $\mathcal{D}_M$ *be stable under multiplication by indicator functions. That is, if* $f \in \mathcal{D}_M$ *then* $\chi_X f \in \mathcal{D}_M$. *Let* $1 \leq p_1 < p_2 \leq \infty$, $1 \leq q_1 < q_2 \leq \infty$, *with* $p_1 \leq q_1$ *and* $p_2 \leq q_2$. *Furthermore, let* $T : \mathcal{D}_M \to \mathscr{F}_N$ *be a sublinear map that is of weak type* (p_1, q_1) *and* (p_2, q_2). *Then, for all* $p \in (p_1, p_2)$, T *is of strong type* (p,q) *with* q *satisfying*

$$\frac{1}{p} = \frac{1-\theta}{p_1} + \frac{\theta}{p_2} \quad and \quad \frac{1}{q} = \frac{1-\theta}{q_1} + \frac{\theta}{q_2}.$$

Proof. We prove the case when $q_1 = p_1, q_2 = p_2 < \infty$ and leave the general case as an exercise.

First, by the weak type (p_i, p_i) hypothesis for $i = 1, 2$ we have constants $C_i > 0$ such that

$$\|Tf\|_{p_i,\infty} \leq C_i\|f\|_{p_i}$$

for all $f \in \mathcal{D}_M \cap \mathrm{L}^{p_i}(M,\mu)$. Fix such an f along with $p \in (p_1, p_2)$ and

$$\int_N |Tf|^p \, d\nu = p \int_0^\infty \nu \{x \in N : |Tf(x)| > \lambda\} \lambda^{p-1} \, d\lambda.$$

Fix $\lambda > 0$ and define

$$g_\lambda(x) = \begin{cases} f(x) & |f(x)| > \lambda \\ 0 & \text{otherwise} \end{cases}$$

and

$$h_\lambda(x) = f(x) - g_\lambda(x).$$

So, $g_\lambda = f\chi_{\{x \in M : |f(x)| > \lambda\}} \in \mathcal{D}_M$ and $h_\lambda = f\chi_{\{x \in M : |f(x)| \leq \lambda\}} \in \mathcal{D}_M$ by the stability hypothesis on $\mathcal{D}_M$.

Now,

$$\int_M |g_\lambda|^{p_1} \, d\mu = \int_{\{x \in M : |f(x)| > \lambda\}} |f|^{p_1} \, d\mu \leq \int_M |f|^{p_1} \left|\frac{f}{\lambda}\right|^{p-p_1} \, d\mu \leq \frac{\|f\|_p^p}{\lambda^{p-p_1}}$$

and by similar calculation,

$$\int_M |h_\lambda|^{p_2} = \int_{\{x \in M : |f(x)| \leq \lambda\}} |f|^{p_2} \, d\mu \leq \lambda^{p_2-p} \int_M |f|^p \, d\mu < \infty.$$

33

Since T is sublinear, for ν-almost all $x \in N$,

$$|Tf(x)| \le |Tg_\lambda(x)| + |Th_\lambda(x)|$$

and so

$$\nu\left\{x \in N : |Tf(x)| > \lambda\right\} \le \nu\left\{x \in N : |Tg_\lambda(x)| > \frac{\lambda}{2}\right\} + \nu\left\{x \in N : |Th_\lambda(x)| > \frac{\lambda}{2}\right\}.$$

We compute,

$$\int_0^\infty \nu\left\{x \in N : |Tg_\lambda(x)| > \frac{\lambda}{2}\right\} \lambda^{p-1} \, d\lambda$$

$$\le \int_0^\infty \frac{C_1^{p_1}}{\left(\frac{\lambda}{2}\right)^{p_1}} \|g_\lambda\|_{p_1}^{p_1} \lambda^{p-1} \, d\lambda$$

$$\le (2C_1)^{p_1} \int_0^\infty \lambda^{p-p_1-1} \int_{\{x \in M : |f(x)| \le \lambda\}} |f|^{p_1} \, d\mu d\lambda$$

$$= (2C_1)^{p_1} \int_M \left(\int_0^{|f|} \lambda^{p-p_1-1} \, d\lambda\right) d\mu$$

$$= \frac{(2C_1)^{p_1}}{p - p_1} \int_M |f|^p \, d\mu.$$

By a similar calculation, but this time integrating from $|f|$ to ∞ with $\lambda^{-\xi}$ for $\xi > 0$, we get

$$\int_0^\infty \nu\left\{x \in N : |Th_\lambda(x)| > \frac{\lambda}{2}\right\} \, d\lambda \le \frac{(2C_2)^{p_2}}{p_2 - p} \int_M |f|^p \, d\mu.$$

Putting these estimates together,

$$\int_N |Tf|^p \le p\left(\frac{(2C_1)^{p_1}}{p - p_1} + \frac{(2C_2)^{p_2}}{p_2 - p}\right) \int_M |f|^p \, d\mu$$

which completes the proof. $\qquad\square$

Remark 4.1.13. *If the definition of g_λ was changed to $g_\lambda = f\chi_{\{x \in M : |f(x)| > a\lambda\}}$ with $a \in \mathbb{R}_+$, then this leads to better bounds by optimising a. In fact, we can get a log convex combination of C_1, C_2. That is:*

$$\|Tf\|_p \le C(p, p_1, p_2) C_1^{1-\theta} C_2^\theta \|f\|_p.$$

For a function $\theta \mapsto h(\theta)$, Log convex here means that $\theta \mapsto \log h(\theta)$ is convex.

4.2 Complex interpolation

We begin by considering a "baby" version of complex interpolation. Let $a \in \mathbb{C}^N$. Then, for $1 \le p < \infty$,

$$\|a\|_p = \left(\sum_{i=1}^N |a_i|^p\right)^{\frac{1}{p}}$$

34

and for $p = \infty$

$$\|a\|_\infty = \sup_{1 < i < N} |a_i|$$

are norms. Let M be an $N \times N$ matrix, then

$$\|M\|_{(\mathcal{L}(\mathbb{C}^N), \|\cdot\|_p)} = \|M\|_{p,p} = \sup_{a \in \mathbb{C}^N, \|a\|_p \neq 0} \frac{\|Ma\|_p}{\|a\|_p}$$

is the associated operator norm. We leave it as an (easy) exercise to verify that

$$\|M\|_{\infty,\infty} = \sup_i \sum_{j=1}^N |m_{ij}|$$

where $M = (m_{ij})$. Then,

$$\|M\|_{1,1} = \|M^*\|_{\infty,\infty} = \sup_j \sum_{i=1}^N |m_{ij}|$$

where we have used the fact that $\|\cdot\|_1$ and $\|\cdot\|_\infty$ are dual norms on $\mathbb{C}^N$. For $1 < p < \infty$, there is no such characterisation of $\|M\|_{p,p}$ but we have the following *Schur's Lemma*.

Proposition 4.2.1 (Schur's Lemma).

$$\|M\|_{p,p} \leq \|M\|_{1,1}^{\frac{1}{p}} \|M\|_{\infty,\infty}^{1 - \frac{1}{p}}.$$

Proof. For matrices, the result follows from the application of Hölder's inequality.

Let $a \in \mathbb{C}^N$ with $\|a\|_p = 1$. Then,

$$
\begin{aligned}
|(Ma)_i| &= \left| \sum_j m_{ij} a_j \right| \\
&\leq \sum_j |m_{ij}| |a_j| \\
&= \sum_j |m_{ij}|^{1 - \frac{1}{p}} |m_{ij}|^{\frac{1}{p}} |a_j| \\
&\leq \left(\sum_j |m_{ij}| \right)^{\frac{1}{p'}} \left(\sum_j |m_{ij}| |a_j|^p \right)^{\frac{1}{p}}
\end{aligned}
$$

and so it follows that

$$\sum_i |(Ma)_i|^p \leq \|M\|_{\infty,\infty}^{\frac{p}{p'}} \sum_i \sum_j |m_{ij}| |a_j|^p \leq \|M\|_{\infty,\infty}^{\frac{p}{p'}} \|M\|_{1,1} \|a\|^p$$

which finishes the proof. $\qquad \square$

This is really a special case of more general interpolation results of the form

$$\|M\|_{q,q} \leq \|M\|_{p,p}^{1-\theta}\|M\|_{r,r}^{\theta}$$

for $1 \leq p < q < r \leq \infty$ and

$$\frac{1}{q} = \frac{1-\theta}{p} + \frac{\theta}{r}.$$

A proof of this statement cannot be accessed easily via elementary techniques since there is no explicit characterisation of $\|M\|_{q,q}$ and $\|M\|_{r,r}$. This was the essential ingredient of the preceding proof. This more general statement comes as a consequence of the powerful complex interpolation method.

We prove this by employing the following 3 lines theorem of Hadamard. First note that in the following statement, $\mathring{S}$ denotes the topological interior of S and $H(\mathring{S})$ denotes the holomorphic functions on $\mathring{S}$.

Lemma 4.2.2 (3 lines Theorem of Hadamard). *Let $S = \{\zeta \in \mathbb{C} : 0 \leq \operatorname{Re}\zeta \leq 1\}$. Let $F \in H(\mathring{S})$ and $F \in C^0(S)$. Further suppose that $\|F\|_\infty < \infty$ on S and let*

$$C_0 = \sup_{t \in \mathbb{R}} |F(\imath t)| \quad and \quad C_1 = \sup_{t \in \mathbb{R}} |F(1 + \imath t)|.$$

Then for $x \in (0,1)$, and $t \in \mathbb{R}$,

$$|F(x + \imath t)| \leq C_0^{1-x}C_1^x.$$

Proof. Assume that $C_0, C_1 > 0$. If not, prove the theorem with $C_0 + \delta, C_1 + \delta$ in place of C_0, C_1 for $\delta > 0$ to conclude that $|F(\zeta)| \leq (C_0 + \delta)^{1-\operatorname{Re}\zeta}(C_1 + \delta)^{\operatorname{Re}\zeta}$ and letting $\delta \to 0$ conclude that

$$|F(\zeta)| = 0 = C_0^{1-\operatorname{Re}\zeta}C_1^{\operatorname{Re}\zeta}$$

for $\zeta \in \mathring{S}$.

Fix $\varepsilon > 0$ and set $G_\varepsilon(\zeta) = F(\zeta)C_0^{\zeta-1}C_1^{-\zeta}e^{\varepsilon\zeta^2}$ for $\zeta \in S$. Then, $G_\varepsilon \in H(\mathring{S}), G_\varepsilon \in C^0(S)$. Now, fix $R > 0$ and let $Q_R = S \cap \{\zeta \in \mathbb{C} : \operatorname{Im}\zeta \leq R\}$. By the maximum principle,

$$\sup_{\zeta \in Q_R} |G_\varepsilon(\zeta)| = \sup_{\zeta \in \partial Q_R} |G_\varepsilon(\zeta)|.$$

We consider each part of ∂Q_R. For $\zeta = \imath t$ with $|t| \leq R$,

$$|G_\varepsilon(\zeta)| = |F(\imath t)| C_0^{-1}e^{\varepsilon(\imath t)^2} \leq 1$$

and for $\zeta = 1 + \imath t$ with $|t| \leq R$,

$$|G_\varepsilon(\zeta)| = |F(1 + \imath t)| C_1^{-1}e^{\operatorname{Re}\,\varepsilon(1-t^2+2\imath t)} \leq e^\varepsilon.$$

Then, for $\zeta = x + \imath R$ with $|t| \leq R$,

$$|G_\varepsilon(\zeta)| \leq |F(x + \imath R)| C_0^{x-1}C_1^x e^{\operatorname{Re}\,\varepsilon(x^2-R^2+2\imath xR)} \leq Ce^{\operatorname{Re}\,\varepsilon(1-R^2)}$$

where

$$C = \left(\sup_{\zeta \in S} |F(\zeta)|\right)\left(\sup_{0 \leq x \leq 1} C_0^{x-1}C_1^x\right)$$

and lastly, when $\zeta = x - \imath R$ with $|t| \leq R$,

$$|G_\varepsilon(\zeta)| \leq C e^{\mathrm{Re}\ \varepsilon(1-R^2)}$$

by the same calculation since $\mathrm{Re}\ \varepsilon(x^2 - R^2 + 2\imath x R) = \varepsilon(x^2 - R^2)$. So, for ε fixed, there exists an R_0 such that whenever $R \geq R_0$,

$$\begin{cases} C e^{-\varepsilon R^2} \leq 1 \\ |G_\varepsilon(x)| \leq e^\varepsilon \\ \forall \zeta \in \mathcal{S}, \ \exists R' \geq R_0 \text{ such that } \zeta \in Q_{R'} \text{ and } |G_\varepsilon(\zeta)| \leq e^\varepsilon \end{cases}$$

and so for all $\zeta \in \mathcal{S}$ and all $\varepsilon > 0$,

$$|F(\zeta)| \leq \left| C_0^{1-\zeta} C_1^\zeta e^{-\varepsilon \zeta^2} e^\varepsilon \right|.$$

The proof is then completed by fixing ζ and letting $\varepsilon \to 0$. $\qquad\square$

Remark 4.2.3. *This lemma can be proved assuming some growth on F for $|\mathrm{Im}\ \zeta| \to \infty$ (rather than F bounded). However, there needs to be some control on the growth.*

Lemma 4.2.4 (Phragmen-Lindelöf)**.** *Let $\Sigma \subset \mathbb{C}$ be the closed subset between the lines $\mathbb{R} \pm \imath \frac{\pi}{2}$ (obtained from the conformal map $\zeta \mapsto \imath \pi(\zeta - \frac{1}{2})$). Suppose $F \in \mathrm{H}(\mathring{\Sigma}), \mathrm{C}^0(\Sigma)$ and assume that F is bounded on the lines $x \pm \imath \frac{\pi}{2}$. Suppose there exists an $A > 0, \beta \in [0,1)$ such that for all $\zeta \in \Sigma$,*

$$|F(\zeta)| \leq \exp(A \exp(\beta\, |\mathrm{Re}\ \zeta|)).$$

Then, $|F(\zeta)| \leq 1$ for $\zeta \in \Sigma$.

Proof. The proof is the same as the proof of the preceding lemma, except $e^{\varepsilon(\imath \zeta)^2} = e^{-\varepsilon \zeta^2}$ term needs to be replaced by something decreasing faster to 0 as $|\mathrm{Re}\ \zeta| \to \infty$. We leave the details as an exercise (or see [Boa54]). $\qquad\square$

We are now in a position to state and prove the powerful complex interpolation theorem of Riesz-Thorin.

Theorem 4.2.5 (Riesz-Thorin)**.** *Let $1 \leq p < r \leq \infty$ and let (M, μ), (N, ν) be σ-finite measure spaces. Let $\mathcal{D}_M$ denote the space of simple, integrable functions on M, and $T : \mathcal{D}_M \to \mathcal{F}_N$ be $\mathbb{C}$-linear. Furthermore, assume that T is of strong type (p,p) and (r,r) with bounds M_p and M_r respectively. Then, T is of strong type (q,q) for any $q \in (p,r)$ with bound M_q satisfying*

$$M_q \leq M_p^{1-\theta} M_r^\theta,$$

where

$$\frac{1}{q} = \frac{1-\theta}{p} + \frac{\theta}{r}$$

with $0 < \theta < 1$.

Before we prove the theorem, we illustrate the following immediate and important corollary.

This is really a special case of more general interpolation results of the form

$$\|M\|_{q,q} \leq \|M\|_{p,p}^{1-\theta} \|M\|_{r,r}^{\theta}$$

for $1 \leq p < q < r \leq \infty$ and

$$\frac{1}{q} = \frac{1-\theta}{p} + \frac{\theta}{r}.$$

A proof of this statement cannot be accessed easily via elementary techniques since there is no explicit characterisation of $\|M\|_{q,q}$ and $\|M\|_{r,r}$. This was the essential ingredient of the preceding proof. This more general statement comes as a consequence of the powerful complex interpolation method.

We prove this by employing the following 3 lines theorem of Hadamard. First note that in the following statement, $\overset{\circ}{\mathcal{S}}$ denotes the topological interior of $\mathcal{S}$ and $H(\overset{\circ}{\mathcal{S}})$ denotes the holomorphic functions on $\overset{\circ}{\mathcal{S}}$.

Lemma 4.2.2 (3 lines Theorem of Hadamard)**.** *Let* $\mathcal{S} = \{\zeta \in \mathbb{C} : 0 \leq \mathrm{Re}\ \zeta \leq 1\}$*. Let* $F \in H(\overset{\circ}{\mathcal{S}})$ *and* $F \in C^0(\mathcal{S})$*. Further suppose that* $\|F\|_{\infty} < \infty$ *on* $\mathcal{S}$ *and let*

$$C_0 = \sup_{t \in \mathbb{R}} |F(\imath t)| \quad \text{and} \quad C_1 = \sup_{t \in \mathbb{R}} |F(1 + \imath t)|.$$

Then for $x \in (0,1)$*, and* $t \in \mathbb{R}$*,*

$$|F(x + \imath t)| \leq C_0^{1-x} C_1^x.$$

Proof. Assume that $C_0, C_1 > 0$. If not, prove the theorem with $C_0 + \delta, C_1 + \delta$ in place of C_0, C_1 for $\delta > 0$ to conclude that $|F(\zeta)| \leq (C_0 + \delta)^{1-\mathrm{Re}\ \zeta} (C_1 + \delta)^{\mathrm{Re}\ \zeta}$ and letting $\delta \to 0$ conclude that

$$|F(\zeta)| = 0 = C_0^{1-\mathrm{Re}\ \zeta} C_1^{\mathrm{Re}\ \zeta}$$

for $\zeta \in \overset{\circ}{\mathcal{S}}$.

Fix $\varepsilon > 0$ and set $G_{\varepsilon}(\zeta) = F(\zeta) C_0^{\zeta-1} C_1^{-\zeta} e^{\varepsilon \zeta^2}$ for $\zeta \in \mathcal{S}$. Then, $G_{\varepsilon} \in H(\overset{\circ}{\mathcal{S}}), G_{\varepsilon} \in C^0(\mathcal{S})$. Now, fix $R > 0$ and let $Q_R = \mathcal{S} \cap \{\zeta \in \mathbb{C} : \mathrm{Im}\ \zeta \leq R\}$. By the maximum principle,

$$\sup_{\zeta \in Q_R} |G_{\varepsilon}(\zeta)| = \sup_{\zeta \in \partial Q_R} |G_{\varepsilon}(\zeta)|.$$

We consider each part of ∂Q_R. For $\zeta = \imath t$ with $|t| \leq R$,

$$|G_{\varepsilon}(\zeta)| = |F(\imath t)| C_0^{-1} e^{\varepsilon(\imath t)^2} \leq 1$$

and for $\zeta = 1 + \imath t$ with $|t| \leq R$,

$$|G_{\varepsilon}(\zeta)| = |F(1 + \imath t)| C_1^{-1} e^{\mathrm{Re}\ \varepsilon(1 - t^2 + 2\imath t)} \leq e^{\varepsilon}.$$

Then, for $\zeta = x + \imath R$ with $|t| \leq R$,

$$|G_{\varepsilon}(\zeta)| \leq |F(x + \imath R)| C_0^{x-1} C_1^x e^{\mathrm{Re}\ \varepsilon(x^2 - R^2 + 2\imath x R)} \leq C e^{\mathrm{Re}\ \varepsilon(1 - R^2)}$$

where

$$C = \left(\sup_{\zeta \in \mathcal{S}} |F(\zeta)| \right) \left(\sup_{0 \leq x \leq 1} C_0^{x-1} C_1^x \right)$$

36

and lastly, when $\zeta = x - \imath R$ with $|t| \leq R$,

$$|G_\varepsilon(\zeta)| \leq C e^{\mathrm{Re}\ \varepsilon(1-R^2)}$$

by the same calculation since $\mathrm{Re}\ \varepsilon(x^2 - R^2 + 2\imath x R) = \varepsilon(x^2 - R^2)$. So, for ε fixed, there exists an R_0 such that whenever $R \geq R_0$,

$$\begin{cases} C e^{-\varepsilon R^2} \leq 1 \\ |G_\varepsilon(x)| \leq e^\varepsilon \\ \forall \zeta \in \mathcal{S},\ \exists R' \geq R_0 \text{ such that } \zeta \in Q_{R'} \text{ and } |G_\varepsilon(\zeta)| \leq e^\varepsilon \end{cases}$$

and so for all $\zeta \in \mathcal{S}$ and all $\varepsilon > 0$,

$$|F(\zeta)| \leq \left| C_0^{1-\zeta} C_1^{\zeta} e^{-\varepsilon \zeta^2} e^\varepsilon \right|.$$

The proof is then completed by fixing ζ and letting $\varepsilon \to 0$. $\qquad\square$

Remark 4.2.3. *This lemma can be proved assuming some growth on F for $|\mathrm{Im}\ \zeta| \to \infty$ (rather than F bounded). However, there needs to be some control on the growth.*

Lemma 4.2.4 (Phragmen-Lindelöf). *Let $\Sigma \subset \mathbb{C}$ be the closed subset between the lines $\mathbb{R} \pm \imath \frac{\pi}{2}$ (obtained from the conformal map $\zeta \mapsto \imath \pi(\zeta - \frac{1}{2})$). Suppose $F \in \mathrm{H}(\mathring{\Sigma}), \mathrm{C}^0(\Sigma)$ and assume that F is bounded on the lines $x \pm \imath \frac{\pi}{2}$. Suppose there exists an $A > 0, \beta \in [0,1)$ such that for all $\zeta \in \Sigma$,*

$$|F(\zeta)| \leq \exp(A \exp(\beta |\mathrm{Re}\ \zeta|)).$$

Then, $|F(\zeta)| \leq 1$ for $\zeta \in \Sigma$.

Proof. The proof is the same as the proof of the preceding lemma, except $e^{\varepsilon(\imath\zeta)^2} = e^{-\varepsilon\zeta^2}$ term needs to be replaced by something decreasing faster to 0 as $|\mathrm{Re}\ \zeta| \to \infty$. We leave the details as an exercise (or see [Boa54]). $\qquad\square$

We are now in a position to state and prove the powerful complex interpolation theorem of Riesz-Thorin.

Theorem 4.2.5 (Riesz-Thorin). *Let $1 \leq p < r \leq \infty$ and let $(M, \mu), (N, \nu)$ be σ-finite measure spaces. Let $\mathcal{D}_M$ denote the space of simple, integrable functions on M, and $T : \mathcal{D}_M \to \mathscr{F}_N$ be $\mathbb{C}$-linear. Furthermore, assume that T is of strong type (p,p) and (r,r) with bounds M_p and M_r respectively. Then, T is of strong type (q,q) for any $q \in (p,r)$ with bound M_q satisfying*

$$M_q \leq M_p^{1-\theta} M_r^{\theta},$$

where

$$\frac{1}{q} = \frac{1-\theta}{p} + \frac{\theta}{r}$$

with $0 < \theta < 1$.

Before we prove the theorem, we illustrate the following immediate and important corollary.

Corollary 4.2.6. *With the above assumptions, T has a continuous extension to a bounded operator $L^q(M, d\mu) \to L^q(N, d\nu)$ for each $q \in (p, r)$.*

Proof. Fix $q \in (p, r)$ and since $q < \infty$, $\mathcal{D}_M$ is dense in $L^q(M, d\mu)$. Then, define the extension by the usual density argument. $\qquad\square$

Proof of Riesz-Thorin. Fix $p < q < r$ and note that $\mathcal{D}_M$ is dense in $L^q(M, d\mu)$, $\mathcal{D}_N$ is dense in $L^{q'}(N, d\nu)$ and by duality, $L^q(N, d\nu)' = L^{q'}(N, d\nu)$. Thus, it suffices to show that

$$A = \sup_{f \in \mathcal{D}_M, \, \|f\|_q = 1, \, g \in \mathcal{D}_N, \, \|g\|_{q'} = 1} \left| \int_N Tf \, g \, d\nu \right| < \infty$$

as this will imply that T is of strong type (q, q) on $\mathcal{D}_M$ and $\|T\| \leq A$.

Fix $f \in \mathcal{D}_M, g \in \mathcal{D}_N$ with $\|f\|_q = \|g\|_{q'} = 1$. So,

$$f = \sum_k \alpha_k \chi_{A_k}$$

where $\alpha_k \in \mathbb{C}, \mu(A_k) < \infty$ and the sum is finite. Similarly,

$$g = \sum_l \beta_l \chi_{B_l},$$

and it follows that

$$\int_N Tf \, g \, d\nu = \sum_k \sum_l \alpha_k \beta_l \int_N T\chi_{A_k} \, \chi_{B_l} \, d\nu.$$

But $\mu(A_k) < \infty$ which implies that $\chi_{A_k} \in L^p(M, d\mu) \cap L^r(M, d\mu)$ and

$$\int_N T\chi_{A_k} \, \chi_{B_l} \, d\nu$$

is well defined. We will construct an $F \in H(\mathring{S}), C^0(S), \|F\|_\infty < \infty$ (where S is defined in Lemma 4.2.2) with

$$F(\theta) = \int_N Tf \, g \, d\nu.$$

Let $\zeta \in \mathbb{C}$ and write

$$f_\zeta = \sum_k |\alpha_k|^{a(\zeta)} \frac{\alpha_k}{|\alpha_k|} \chi_{A_k}$$

where

$$a(\zeta) = \frac{q}{p}(1 - \zeta) + \frac{q}{r}\zeta.$$

Similarly,

$$g_\zeta = \sum_l |\beta_l|^{a(\zeta)} \frac{\beta_l}{|\beta_l|} \chi_{B_l}$$

and

$$a(\zeta) = \frac{q'}{p'}(1 - \zeta) + \frac{q'}{r'}\zeta.$$

38

Note that $a(\theta) = 1$ if and only if $b(\theta) = 1$, and $f = f_\theta$ and $g = g_\theta$. Set

$$F(\zeta) = \int_N Tf_\zeta \, g_\zeta \, d\nu$$

and note that it is well defined for each ζ by the same reasoning as previously for f and g in place of f_ζ and g_ζ. Furthermore,

$$F(\zeta) = \sum_k \sum_l |\alpha_k|^{a(\zeta)} |\beta_k|^{b(\zeta)} \frac{\alpha_k}{|\alpha_k|} \frac{\beta_l}{|\beta_l|} \int_N T\chi_{A_k} \, \chi_{B_l} \, d\nu$$

so $F \in H(\mathbb{C}) \subset H(\mathring{S}), C^0(S)$ and

$$F(\theta) = \int_N Tf \, g \, d\nu.$$

In order to apply Lemma 4.2.2 we estimate $\sup_{t \in \mathbb{R}} |F(it)|, \sup_{t \in \mathbb{R}} |F(1 + it)|$. So, by the strong type (p, p) property of T,

$$\sup_{t \in \mathbb{R}} |F(it)| = \sup_{t \in \mathbb{R}} \left| \int_N Tf_{it} \, g_{it} \, d\nu \right|$$
$$\leq \sup_{t \in \mathbb{R}} \|Tf_{it}\|_p \|g_{it}\|_{p'}$$
$$\leq M_p \sup_{t \in \mathbb{R}} \|f_{it}\|_p \|g_{it}\|_{p'}.$$

Using the fact that A_k are mutually disjoint, we can write

$$\|f_{it}\|_p^p = \sum_k \left| |\alpha_k|^{a(it)} \right|^p \mu(A_k)$$

and since $a(it) = \frac{q}{p} + it\left(\frac{q}{r} - \frac{q}{p}\right)$,

$$\|f_{it}\|_p^p = \sum_k |\alpha_k|^q \, \mu(A_k) = \|f\|_q^q.$$

By a similar calculation, $\|g_{it}\|_{p'}^{p'} = \|g\|_{q'}^{q'}$ if $p' < \infty$ and $\|g_{it}\|_\infty = 1 = \|g\|_{q'}^{q'}$ if $p' = \infty$.

This shows that $\sup_{it} |F(it)| \leq M_p$. An identical calculation, using the strong type (r, r) property of T gives $\sup_{it} |F(1 + it)| \leq M_r$. Then, set $C_0 = M_p$ and $C_1 = M_r$ and we invoke Lemma 4.2.2 to find

$$|F(\zeta)| \leq M_p^{1 - \operatorname{Re} \zeta} M_r^{\operatorname{Re} \zeta}.$$

The proof is then complete by setting $\zeta = \theta$. $\qquad\square$

Exercise 4.2.7. *Assume the hypothesis of the previous theorem, but here assume that T is of strong type (p_1, p_2) and strong type (r_1, r_2) where $1 \leq p_1, r_1 \leq \infty$ and $1 \leq p_2, r_2 \leq \infty$. Then, T is of strong type (q_1, q_2) where*

$$\frac{1}{q_1} = \frac{1 - \theta}{p_1} + \frac{\theta}{r_1} \qquad and \qquad \frac{1}{q_2} = \frac{1 - \theta}{p_2} + \frac{\theta}{r_2}$$

where $0 < \theta < 1$.

Remark 4.2.8. *Notice that in the previous exercise, we do not require $p_1 < p_2$ and $r_1 < r_2$ as we did in the case of Real interpolation (Theorem 4.1.12).*

Chapter 5

Bounded Mean Oscillation

In this chapter, the framework we work within is $\mathbb{R}^n$ with the metric $d = d_\infty$ and Lebesgue measure $\mathscr{L}$. By $Q(x, r)$, we always denote a ball with respect to d of radius r centred at x. That is, $Q = Q(x, r)$ represents an arbitrary cube in $\mathbb{R}^n$.

We remark that for the theory, there is nothing special about $\mathbb{R}^n$ and d_∞. It is just convenient to work in this setting. The following material could be defined and studied similarly on spaces of *homogeneous type*.

5.1 Construction and properties of BMO

We introduce some notation. Let $\mathrm{m}_X f$ denote the mean of the function f on the set X. That is

$$\mathrm{m}_X f = \frac{1}{\mathscr{L}(X)} \int_X f \, d\mathscr{L}.$$

Definition 5.1.1 (Bounded Mean Oscillation (BMO)). *Let $f \in \mathrm{L}^1_{\mathrm{loc}}(\mathbb{R}^n)$, real or complex valued. We say that f has* bounded mean oscillation *if*

$$\|f\|_* = \sup_Q \mathrm{m}_Q |f - \mathrm{m}_Q f| = \sup_Q \frac{1}{\mathscr{L}(Q)} \int_Q |f - \mathrm{m}_Q f| \, d\mathscr{L} < \infty.$$

Here Q is an arbitrary cube in $\mathbb{R}^n$. We define

$$\mathrm{BMO} = \left\{ f \in \mathrm{L}^1_{\mathrm{loc}}(\mathbb{R}^n) : \|f\|_* < \infty \right\}.$$

Proposition 5.1.2 (Properties of BMO). *(i) $\mathrm{L}^\infty(\mathbb{R}^n) \subset \mathrm{BMO}$ and $\|f\|_* \leq 2\|f\|_\infty$.*

(ii) BMO is a linear space (over $\mathbb{K}$), and $\|\cdot\|_*$ is a semi-norm. That is,

$$\|f + g\|_* \leq \|f\|_* + \|g\|_*,$$

and

$$\|\lambda f\|_* = |\lambda| \, \|f\|_*$$

for $f, g \in$ BMO and $\lambda \in \mathbb{K}$. Furthermore, $\|f\|_* = 0$ if and only if f constant almost everywhere $x \in \mathbb{R}^n$.

(iii) BMO$/\mathbb{K}$ *is a normed space with norm*

$$\|f + \mathbb{K}\| = \|f\|_*,$$

making BMO$/\mathbb{K}$ *a Banach space. BMO convergence is often called* convergence *modulo constant.*

(iv) For $f \in L^\infty(\mathbb{R}^n)$, $x_0 \in \mathbb{R}^n$ and $t > 0$ the function defined by

$$f_{t,x_0}(x) = f\left(\frac{x - x_0}{t}\right) \in L^\infty(\mathbb{R}^n).$$

Similarly, for $f \in$ BMO, $f_{t,x_0} \in$ BMO and $\|f_{t,x_0}\|_* = \|f\|_*$.

Proof. We prove that $\|f\|_* = 0$ if and only if f constant almost everywhere $x \in \mathbb{R}^n$ (ii) and leave the rest as an exercise.

Note that the "if" direction is trivial. To prove the "only if" direction, assume that $\|f\|_* = 0$. Then, for every cube $Q \subset \mathbb{R}^n$, $f = \mathrm{m}_Q f$ almost everywhere $x \in Q$. Let Q_j be an exhaustion of $\mathbb{R}^n$ by increasing cubes. That is, let $Q_j = [-2^j, 2^j]^n$ for $j = 0, 1, 2, \ldots$. Then, $f = \mathrm{m}_{Q_j} f$ almost everywhere on Q_j and hence $\mathrm{m}_{Q_0} f = \mathrm{m}_{Q_j} f$ for all j. Letting $j \to \infty$ we establish that f is constant almost everywhere on $\mathbb{R}^n$. $\qquad\square$

Exercise 5.1.3. *1. Let $A \in GL_n(\mathbb{K})$ and $x_0 \in \mathbb{R}^n$. Write $f_{A,x_0}(x) = f(Ax - x_0)$. Show that if $f \in$ BMO then $f_{A,x_0} \in$ BMO and*

$$\|f_{A,x_0}\|_* \leq 2\|A\|_{\infty,\infty}^n \det A^{-1} \|f\|_*.$$

(Hint: Use the next Lemma). This exercise illustrates that we can indeed change the shape of cubes.

2. Show that

$$\|f\|_*' = \sup_B \mathrm{m}_B(|f - \mathrm{m}_B f|) < \infty$$

where B are Euclidean balls defines an equivalent semi-norm on BMO.

Lemma 5.1.4. *Let $f \in L^1_{\mathrm{loc}}(\mathbb{R}^n)$ and suppose there exists a $C > 0$ such that for all cubes Q, there exists a $c_Q \in \mathbb{K}$ such that $\mathrm{m}_Q |f - c_Q| \leq C$. Then, $f \in$ BMO and $\|f\|_* \leq 2C$.*

Proof. We write

$$f - \mathrm{m}_Q f = f - c_Q + c_Q - \mathrm{m}_Q f = f - c_Q + \mathrm{m}_Q(c_Q - f)$$

and it follows that

$$\mathrm{m}_Q |f - \mathrm{m}_Q f| \leq \mathrm{m}_Q |f - c_Q| + \mathrm{m}_Q |c_Q - f| \leq 2\,\mathrm{m}_Q |f - c_Q| \leq 2C.$$

$\qquad\square$

Example 5.1.5 $(\ln|x| \in \mathrm{BMO})$. *In particular, this implies that* $\mathrm{L}^\infty(\mathbb{R}^n) \subsetneq \mathrm{BMO}$.

We show that this is true for $n = 1$. *Let* $f(x) = \ln|x|$. *For* $t > 0$,

$$f_t(x) = \ln\left|\frac{x}{t}\right| = \ln|x| - \ln|t| = f(x) + c.$$

So, for I *an interval of length* t,

$$\mathrm{m}_I \left|f - \mathrm{m}_I f\right| = \mathrm{m}_I \left|f_t - \mathrm{m}_I f_t\right|.$$

By change of variables, $y = \frac{x}{t} \in J$ *where* J *is of unit length and*

$$\mathrm{m}_I \left|f_t - \mathrm{m}_I f_t\right| = \mathrm{m}_J \left|f - \mathrm{m}_J f\right|$$

and so we are justified in assuming that I *has unit length. So, let* $I = [x_0 - \frac{1}{2}, x_0 + \frac{1}{2}]$ *and by symmetry, assume* $x_0 \geq 0$.

Consider the case $0 \leq x_0 \leq 3$. *Then,*

$$\mathrm{m}_I |f| = \int_I |f(x)|\ dx \leq \int_{-\frac{1}{2}}^{\frac{7}{2}} |\ln|x||\ dx < \infty.$$

Now, suppose $x_0 \geq 3$. *Then, for* $x \in [x_0 - \frac{1}{2}, x_0 + \frac{1}{2}]$,

$$\left|\ln(x) - \ln\left(x_0 - \frac{1}{2}\right)\right| = \ln(x) - \ln\left(x_0 - \frac{1}{2}\right) = \int_{x_0 - \frac{1}{2}}^x \frac{1}{t}\ dt \leq \int_{x_0 - \frac{1}{2}}^x \frac{2}{5}\ dt$$

$$\leq \frac{2}{5}\left(x - x_0 + \frac{1}{2}\right) \leq \frac{2}{5}\left(x_0 + \frac{1}{2} - x_0 + \frac{1}{2}\right) \leq \frac{2}{5}$$

and by setting $C_I = \ln(x_0 - \frac{1}{2})$, $C = \frac{2}{5}$,

$$\fint_{x_0 - \frac{1}{2}}^{x_0 + \frac{1}{2}} |\ln(x) - C_I|\ d\mathscr{L}(x) = \int_{x_0 - \frac{1}{2}}^{x_0 + \frac{1}{2}} |\ln(x) - C_I|\ d\mathscr{L}(x) \leq \frac{2}{5} = C.$$

Let $C_I = 0$ *whenever* $0 \leq x_0 < 3$ *and let*

$$C' = \max\left(C, \int_{-\frac{1}{2}}^{\frac{7}{2}} |\ln|x||\ d\mathscr{L}(x)\right).$$

Then, $\mathrm{m}_I |f - C_I| \leq C'$ *and the proof is complete by applying Lemma 5.1.4.*

The following proposition highlights an important feature of L^∞ which is missing in BMO.

Proposition 5.1.6. *Let*

$$f(x) = \begin{cases} \ln(x) & x > 0 \\ 0 & x \leq 0 \end{cases}.$$

Then, $f \notin \mathrm{BMO}$ *and consequently,* BMO *is not stable under multiplication by indicator functions.*

Proof. Let $I = [-\varepsilon, \varepsilon]$ for $\varepsilon > 0$ and small. We compute,

$$\mathrm{m}_I |f - \mathrm{m}_I f| = \frac{1}{2c} \int_0^\varepsilon |\ln|x| - \mathrm{m}_I f| \; d\mathscr{L}(x) + \frac{1}{2\varepsilon} \int_{-\varepsilon}^0 |\mathrm{m}_I f| \; d\mathscr{L}(x).$$

Then,

$$\begin{aligned}
\frac{1}{2\varepsilon} \int_{-\varepsilon}^0 |\mathrm{m}_I f| \; d\mathscr{L}(x) &= \frac{1}{2\varepsilon} \int_{-\varepsilon}^0 \left| \frac{1}{2\varepsilon} \int_0^\varepsilon \ln|x| \; d\mathscr{L}x \right| \; d\mathscr{L}(y) \\
&= \frac{1}{2} \frac{1}{2\varepsilon} |\varepsilon \ln \varepsilon - \varepsilon| \\
&= \frac{1}{4} |\ln \varepsilon - 1|.
\end{aligned}$$

But this is *not* bounded as $\varepsilon \to 0$. $\qquad\qquad\qquad\qquad\qquad\qquad\qquad\qquad\qquad\qquad$ $\square$

Proposition 5.1.7 (Further properties of BMO). *(i) Whenever $f \in$ BMO, then $|f| \in$ BMO and $\| |f| \|_* \leq 2\|f\|_*$.*

(ii) Suppose $f, g \in$ BMO are real valued. Then, $f_+, f_-, \max(f, g), \min(f, g) \in$ BMO. Furthermore,

$$\| \max(f, g) \|_*, \| \min(f, g) \|_* \leq \frac{3}{2} (\|f\|_* + \|g\|_*).$$

(iii) Let $f \in$ BMO real valued. Then we have the following Approximation by truncation. Let

$$f_N(x) = \begin{cases} N & f(x) > N \\ f(x) & -N \leq f(x) \leq N \\ -N & f(x) < N \end{cases}$$

for $N \in \mathbb{R}_+$. Then, $f_N \in \mathrm{L}^\infty(\mathbb{R}^n)$, $\|f_N\|_* \leq 2\|f\|_*$ and $f_N \to f$ almost everywhere in $\mathbb{R}^n$.

(iv) Assume f is complex valued. Then $f \in$ BMO if and only if $\mathrm{Im}\, f, \mathrm{Re}\, f \in$ BMO and

$$\|\mathrm{Im}\, f\|_*, \|\mathrm{Re}\, f\|_* \leq \|f\|_* \leq \|\mathrm{Im}\, f\|_* + \|\mathrm{Re}\, f\|_*.$$

Proof. (i) Let $C_Q = |\mathrm{m}_Q f|$. Then, $||f| - C_Q| \leq |f - \mathrm{m}_Q f|$ and so $\mathrm{m}_Q ||f| - C_Q| \leq \mathrm{m}_Q |f - \mathrm{m}_Q f| \leq \|f\|_*$. Then, apply Lemma 5.1.4.

(ii) Apply (i). Exercise.

(iii) Pick Q a cube, and let $x, y \in Q$. Then, $|f_N(x) - f_N(y)| \leq |f(x) - f(y)|$ and

$$f_N(x) - \mathrm{m}_Q f_N = \fint_Q (f_N(x) - f_N(y)) \; d\mathscr{L}(y).$$

So,

$$\begin{aligned}
\fint_Q |f_N(x) - \mathrm{m}_Q f_N| &\leq \fint_Q \fint_Q |f_N(x) - f_N(y)| \; d\mathscr{L}(x) d\mathscr{L}(y) \\
&\leq \fint_Q \fint_Q |f(x) - \mathrm{m}_Q f + \mathrm{m}_Q f - f(y)| \; d\mathscr{L}(x) d\mathscr{L}(y) \\
&\leq \fint_Q \fint_Q |f(x) - \mathrm{m}_Q f| + |\mathrm{m}_Q f - f(y)| \; d\mathscr{L}(x) d\mathscr{L}(y) \\
&\leq 2\|f\|_*.
\end{aligned}$$

43

(iv) Exercise.

$\square$

Exercise 5.1.8. *Find a function $f \in L^1_{loc}(\mathbb{R}^n)$ with $|f| \in$ BMO, but $f \notin$ BMO.*

Proposition 5.1.9. *Let Q, R be cubes with $Q \subset R$. Let $f \in$ BMO. Then,*

$$|m_Q f - m_R f| \leq \frac{\mathscr{L}(R)}{\mathscr{L}(Q)} \|f\|_*.$$

Proof. We compute,

$$
\begin{aligned}
|m_Q f - m_R f| &\leq m_Q |f - m_R f| \\
&= \frac{1}{\mathscr{L}(Q)} \int_Q |f - m_R f| \\
&\leq \frac{1}{\mathscr{L}(Q)} \int_R |f - m_R f| \\
&\leq \frac{\mathscr{L}(R)}{\mathscr{L}(Q)} \left(\frac{1}{\mathscr{L}(R)} \right) \int_R |f - m_R f| \\
&\leq \frac{\mathscr{L}(R)}{\mathscr{L}(Q)} \|f\|_*.
\end{aligned}
$$

$\square$

Corollary 5.1.10. *(i) Suppose that $\mathscr{L}(R) \leq 2\mathscr{L}(Q)$ with $Q \subset R$. Then $|m_R f - m_Q f| \leq 2\|f\|_*$.*

(ii) Suppose that Q, R are arbitrary cubes (not necessarily $Q \subset R$). Then, $|m_R f - m_Q f| \leq C_n \|f\|_* \, \rho(Q, R)$ where

$$\rho(Q, R) = \ln \left(2 + \frac{\mathscr{L}(R)}{\mathscr{L}(Q)} + \frac{\mathscr{L}(Q)}{\mathscr{L}(R)} + \frac{\text{dist}(Q, R)}{(\mathscr{L}(Q) \wedge \mathscr{L}(R))^{\frac{1}{n}}} \right).$$

Proof. We leave the proof as an exercise but note that the proof of (i) is easy. The proof of (ii) requires a "telescoping argument."

$\square$

5.2 John-Nirenberg inequality

BMO was invented by John-Nirenberg for use in partial differential equations.

Theorem 5.2.1 (John-Nirenberg inequality). *There exist constants $C = C(n) \geq 0$ and $\alpha = \alpha(n) > 0$ depending on dimension such that for all $f \in$ BMO with $\|f\|_* \neq 0$ and for all cubes Q and $\lambda > 0$,*

$$\mathscr{L}\{x \in Q : |f(x) - m_Q f| > \lambda\} \leq C e^{-\alpha \frac{\lambda}{\|f\|_*}} \mathscr{L}(Q)$$

Remark 5.2.2 (On the exponential decay). *Note that by the definition of the* BMO *norm combined with Tchebitchev-Markov inequality (Proposition 4.1.4), we get decay in $\frac{1}{\lambda}$ since*

$$\mathscr{L}\left\{x \in Q : |f(x) - \mathrm{m}_Q f| > \lambda\right\}$$

$$\leq \frac{1}{\lambda}\int_{\{x \in Q : |f(x) - \mathrm{m}_Q f| > \lambda\}} |f - \mathrm{m}_Q f| \; d\mathscr{L} \leq \frac{1}{\lambda}\|f\|_*\mathscr{L}(Q).$$

It is natural to ask the question why we get extra gain into exponential decay. The reason is as follows. The expression above is for a single cube. But the exponential decay comes from the fact that we have scale and translation invariant estimates. This is typical in harmonic analysis.

Proof of the John-Nirenberg inequality. First, note that it is enough to assume that $Q = Q_0 = [0,1)^n$ by the scale and translation invariance of BMO. Furthermore, we can assume that $\mathrm{m}_{Q_0} f = 0$ since $\|f\|_* = \|f - \mathrm{m}_Q f\|_*$. By multiplying by a constant coupled with the fact that $\|\cdot\|_*$ is a semi-norm, we need to only consider $\|f\|_* = 1$.

Let $\mathcal{F}_\lambda = \{x \in Q : |f(x)| > \lambda\}$. We show that $\mathscr{L}(\mathcal{F}_\lambda) \leq Ce^{-\alpha\lambda}$. We prove this for f_N (the truncation of f) and let $N \to \infty$ to establish the claim via the monotone convergence theorem. So, without loss of generality, assume that $f \in \mathrm{L}^\infty(\mathbb{R}^n)$.

Consider the case when $\lambda > 1$. As before, let $\mathscr{D}(Q)$ denote the dyadic subcubes of Q. Let $\mathcal{E}_\lambda = \{x \in Q : \mathcal{M}^{\mathscr{D}}f(x) > \lambda\}$ and we have that $\mathcal{F}_\lambda \subset \mathcal{E}_\lambda$ up to a set of null measure (since $|f| \leq \mathcal{M}^{\mathscr{D}}f$ almost everywhere). Hence, $\mathscr{L}(\mathcal{F}_\lambda) \leq \mathscr{L}(\mathcal{E}_\lambda)$ and we show that $\mathscr{L}(\mathcal{E}_\lambda) \leq Ce^{-\alpha\lambda}$. Then, by definition

$$\sup_{\mathcal{Q}\in\mathscr{D}(Q),\; \mathcal{Q}\ni x} \mathrm{m}_\mathcal{Q}|f| = \mathcal{M}^{\mathscr{D}}f(x)$$

and

$$\mathrm{m}_Q|f| = \mathrm{m}_Q|f - \mathrm{m}_Q f| \leq \|f\|_* = 1.$$

Coupled with this and the assumption that $\lambda > 1$, we have that $\mathcal{E}_\lambda \subsetneq Q$. Let $\mathscr{C} = \{\mathcal{Q}_{i,\lambda}\}$ be the maximal collection of subcubes $\mathcal{Q} \in \mathscr{D}(Q)$ such that $\mathcal{E}_\lambda = \sqcup \mathcal{Q}_{i,\lambda}$. So,

$$\mathrm{m}_{\mathcal{Q}_{i,\lambda}}|f| > \lambda \qquad \text{and} \qquad \mathrm{m}_{\widehat{\mathcal{Q}_{i,\lambda}}}|f| \leq \lambda$$

with $\widehat{\mathcal{Q}_{i,\lambda}} \subset Q$.

For each i, we estimate $\left|\mathrm{m}_{\mathcal{Q}_{i,\lambda}} f - \mathrm{m}_{\widehat{\mathcal{Q}_{i,\lambda}}} f\right|$. Let $\{R_k\}_{k=0}^n$ be a set of cubes such that

$$\mathcal{Q}_{i,\lambda} = R_0 \subset R_1 \subset \cdots \subset R_n = \widehat{\mathcal{Q}_{i,\lambda}}$$

and

$$\frac{\mathscr{L}(R_{k+1})}{\mathscr{L}(R_k)} \leq 2.$$

It is trivial that such a collection exists. By this we have that for all k,

$$\left|\mathrm{m}_{R_{k+1}} f - \mathrm{m}_{R_k} f\right| \leq 2\|f\|_* = 2$$

and summing over k,

$$\left| m_{Q_{i,\lambda}} f - m_{\widehat{Q_{i,\lambda}}} f \right| \leq 2n.$$

Therefore,

$$\left| m_{Q_{i,\lambda}} f \right| \leq 2n + \left| m_{\widehat{Q_{i,\lambda}}} f \right| \leq 2n + \lambda.$$

Now, pick a $\delta > 2n + 1$ to be chosen later. Let $\mathscr{C}' = \{Q_{j,\lambda+\delta}\}$ be the maximal disjoint covering for $\mathcal{E}_{\lambda+\delta}$. Then, $\mathcal{E}_{\lambda+\delta} \subset \mathcal{E}_\lambda$ and for each j, there exists a unique i such that $Q_{j,\lambda+\delta} \subset Q_{i,\lambda}$. Fix i, and estimate:

$$
\begin{aligned}
\mathscr{L}(\mathcal{E}_{\lambda+\delta} \cap Q_{i,\lambda}) &= \mathscr{L}\left(\sqcup_{\{j: Q_{j,\lambda+\delta} \subset Q_{i,\lambda}\}} Q_{j,\lambda+\delta} \right) \\
&= \sum_{\{j: Q_{j,\lambda+\delta} \subset Q_{i,\lambda}\}} \mathscr{L}(Q_{j,\lambda+\delta}) \\
&\leq \frac{1}{\lambda+\delta} \sum_{\{j: Q_{j,\lambda+\delta} \subset Q_{i,\lambda}\}} \int_{Q_{j,\lambda+\delta}} |f| \, d\mathscr{L} \\
&\leq \frac{1}{\lambda+\delta} \int_{\mathcal{E}_{\lambda+\delta} \cap Q_{i,\lambda}} |f| \, d\mathscr{L} \\
&\leq \frac{1}{\lambda+\delta} \int_{Q_{i,\lambda}} \left| f - m_{Q_{i,\lambda}f} \right| \, d\mathscr{L} + \frac{1}{\lambda+\delta} \left| m_{Q_{i,\lambda}} f \right| \mathscr{L}(\mathcal{E}_{\lambda+\delta} \cap Q_{i,\lambda}) \\
&\leq \frac{1}{\lambda+\delta} |Q_{i,\lambda}| \|f\|_* + \frac{2n+\lambda}{\lambda+\delta} \mathscr{L}(\mathcal{E}_{\lambda+\delta} \cap Q_{i,\lambda}).
\end{aligned}
$$

Therefore,

$$\mathscr{L}(\mathcal{E}_{\lambda+\delta} \cap Q_{i,\lambda}) \leq \frac{1}{\delta - 2n} \mathscr{L}(Q_{i,\lambda})$$

and summing over i,

$$\mathscr{L}(\mathcal{E}_{\lambda+\delta}) \leq \frac{1}{\delta - 2n} \mathscr{L}(\mathcal{E}_\lambda).$$

Now, fix δ such that $1 < \delta - 2n$ and observe that

$$\mathscr{L}(\mathcal{E}_{\lambda+k\delta}) \leq \left(\frac{1}{\delta - 2n} \right)^k \mathscr{L}(\mathcal{E}_\lambda)$$

for $k \in \mathbb{N}$. If $\lambda \geq 2$, then there exists a unique $k \in \mathbb{N}$ such that $2 + k\delta \leq \lambda < 2 + (k+1)\delta$ and so it follows that

$$\mathscr{L}(\mathcal{E}_\lambda) \leq \mathscr{L}(\mathcal{E}_{2+2k\delta}) \leq e^{-k \ln(\delta - 2n)} \mathscr{L}(\mathcal{E}_2) \leq e^{-k \ln(\delta - 2n)} \mathscr{L}(Q) \leq e^{-k \ln(\delta - 2n)} \leq C e^{-\alpha\lambda}.$$

For the case $0 < \lambda \leq 2$,

$$\mathscr{L}(\mathcal{F}_\lambda) \leq \mathscr{L}(Q) \leq e^{\alpha 2} \cdot e^{-\alpha 2} \leq C e^{-\alpha\lambda}$$

and the proof is complete. $\qquad\square$

Definition 5.2.3 (BMO$_p$). *For $f \in L^p_{loc}(\mathbb{R}^n)$, define*

$$\|f\|_{*,p} = \sup_Q (m_Q |f - m_Q f|^p)^{\frac{1}{p}}$$

and define

$$\mathrm{BMO}_p = \left\{ f \in L^p_{loc}(\mathbb{R}^n) : \|f\|_{*,p} < \infty \right\}.$$

Remark 5.2.4. *Note that $\mathrm{BMO}_p \subset \mathrm{BMO}$ and $\|f\|_* \lesssim \|f\|_{*,p}$.*

Corollary 5.2.5. *For all $1 < p < \infty$, $\mathrm{BMO}_p = \mathrm{BMO}$ and $\|f\|_{*,p} \simeq \|f\|_*$.*

Proof. Fix a cube Q and $f \in \mathrm{BMO}$. Then,

$$\int_Q |f - m_Q f|^p \, d\mathscr{L} = \int_0^\infty p\lambda^{p-1} \mathscr{L}\left\{x \in Q : |f(x) - m_Q f| > \lambda\right\} \, d\lambda$$

$$\leq \left(p \int_0^\infty \lambda^{p-1} C e^{-\alpha \frac{\lambda}{\|f\|_*}} \, d\lambda \right) \mathscr{L}(Q)$$

and noting that

$$p \int_0^\infty \lambda^{p-1} C e^{-\alpha \frac{\lambda}{\|f\|_*}} \, d\lambda \leq C\|f\|_*^p$$

completes the proof. $\qquad\square$

Exercise 5.2.6. *For $f \in \mathrm{BMO}$, there exists a $\beta > 0$ such that*

$$\sup_Q \fint_Q \exp(\beta |f - m_Q f|) \, d\mathscr{L} < \infty.$$

5.3   Good λ inequalities and sharp maximal functions

We introduce the following variants on centred and uncentred maximal function. They are constructed using arbitrary cubes rather than balls.

Definition 5.3.1 (Cubic maximal functions). *For $f \in L^1_{loc}(\mathbb{R}^n)$, define the centred cubic maximal function:*

$$\mathcal{M}^\square f(x) = \sup_{r>0} \fint_{Q(x,r)} |f(y)| \, d\mathscr{L}(y)$$

and the uncentred cubic maximal function:

$$\mathcal{M}^{\square'} f(x) = \sup_{Q \ni x} \fint_Q |f(y)| \, d\mathscr{L}(y).$$

Proposition 5.3.2. *There exist $C_1, C_2 > 0$ such that*

$$C_1 \mathcal{M}_{\mathscr{L}} f \leq \mathcal{M}^\square f \leq C_2 \mathcal{M}_{\mathscr{L}} f$$

and

$$C_1 \mathcal{M}'_{\mathscr{L}} f \leq \mathcal{M}^{\square'} f \leq C_2 \mathcal{M}'_{\mathscr{L}} f$$

Proof. The proof follows easily noting that there exist constants A and B such that for every cube Q, there exist balls B_1, B_2 with the same centre such that $B_1 \subset Q \subset B_2$ and $A\mathscr{L}(B_1) = \mathscr{L}(Q) = B\mathscr{L}(B_2)$. $\qquad\square$

Remark 5.3.3. *In particular, this means that we can simply substitute $\mathcal{M}^\square$ and $\mathcal{M}^{\square'}$ in place of $\mathcal{M}_{\mathscr{L}}$ and $\mathcal{M}'_{\mathscr{L}}$ in Theorems and obtain the same conclusions.*

We now introduce a new type of maximal function which will be the primary tool of this section.

Definition 5.3.4 (Sharp maximal function). *For $f \in \mathrm{L}^1_{\mathrm{loc}}(\mathbb{R}^n)$ and $x \in \mathbb{R}^n$, define*

$$\mathcal{M}^\sharp f(x) = \sup_{Q \ni x} \mathrm{m}_Q |f - \mathrm{m}_Q\, f|\,.$$

Remark 5.3.5. *(i) $f \in \mathrm{BMO}$ if and only if $\mathcal{M}^\sharp f \in \mathrm{L}^\infty(\mathbb{R}^n)$. Furthermore, $\|f\|_* = \|\mathcal{M}^\sharp f\|_\infty$.*

(ii) $\mathcal{M}^\sharp f \leq 2\mathcal{M}^{\square'} f$.

In particular (ii) means that if $f \in \mathrm{L}^p(\mathbb{R}^n)$ with $1 < p < \infty$, then $\mathcal{M}^\sharp f \in \mathrm{L}^p(\mathbb{R}^n)$.

It is natural to ask whether there is a converse to (ii) in the previous remark. There is no pointwise inequality - consider f constant. This is also true for L^p functions (see [Ste93]). The only hope is to prove $\|\mathcal{M}^{\square'} f\|_p \lesssim \|\mathcal{M}^\sharp f\|_p$ for $f \in \mathrm{L}^p(\mathbb{R}^n)$.

Good λ inequalities (originally from probability theory) help us to establish such a bound. These are distributional inequalities of the following type:

Definition 5.3.6 (Good λ inequality). *A good λ inequality is of the form*

$$\mathscr{L}\left\{x \in \mathbb{R}^n : |f(x)| > \kappa\lambda,\ |g(x)| \leq \gamma\lambda\right\} \leq \varepsilon(\kappa, \gamma)\, \mathscr{L}\left\{x \in \mathbb{R}^n : |f(x)| > \lambda\right\} \qquad (\mathcal{I}_\lambda)$$

where f, g are measurable, $\lambda > 0$, $\kappa > 1$, $\gamma \in (0,1)$ and $\varepsilon(\kappa, \gamma) > 0$.

Proposition 5.3.7. *Suppose there exists a $p_0 \in (0, \infty)$ such that $\|f\|_{p_0} < \infty$ and assume $(\mathcal{I}_\lambda)$ holds for all $\lambda > 0$. Then, for all $p \in [p_0, \infty)$ satisfying*

$$\frac{1}{C_p} = \sup_{\kappa > 1,\ \gamma \in (0,1)} (1 - \kappa^p \varepsilon(\kappa, \gamma)) > 0$$

we have

$$\|f\|_p \leq (2C_p)^{\frac{1}{p}} \frac{\kappa}{\gamma} \|g\|_p$$

for some $\kappa > 1$ and $\gamma < 1$.

Proof. Let $f_N = f\chi_{\{x \in \mathbb{R}^n : |f(x)| \leq N\}}$ and set $I_N = \int_0^N \lambda^{p-1} \mathscr{L}\left\{x \in \mathbb{R}^n : |f(x)| > \lambda\right\}\, d\lambda$. Then note that

$$\int |f_N|^p = p \int_0^\infty \lambda^{p-1} \mathscr{L}\left\{x \in \mathbb{R}^n : |f_N(x)| > \lambda\right\}\, d\lambda \leq pI_N.$$

48

By a change of variable,

$$I_N = \kappa^p \int_0^{\frac{N}{\kappa}} \lambda^{p-1} \mathscr{L} \left\{ x \subset \mathbb{R}^n : |f(x)| > \kappa\lambda \right\} \, d\lambda.$$

Also,

$$\{x \in \mathbb{R}^n : |f(x)| > \kappa\lambda\} \subset \{x \in \mathbb{R}^n : |f(x)| > \kappa\lambda, \; |g(x)| \le \gamma\lambda\} \cup \{x \in \mathbb{R}^n : |g(x)| > \gamma\lambda\}$$

and so it follows that

$$\begin{aligned}
\mathscr{L} \left\{ x \in \mathbb{R}^n : |f(x)| > \kappa\lambda \right\} &\le \mathscr{L} \left\{ x \in \mathbb{R}^n : |f(x)| > \kappa\lambda, \; |g(x)| \le \gamma\lambda \right\} \\
&\quad + \mathscr{L} \left\{ x \in \mathbb{R}^n : |g(x)| > \gamma\lambda \right\} \\
&\le \varepsilon(\kappa,\gamma) \mathscr{L} \left\{ x \in \mathbb{R}^n : |f(x)| > \lambda \right\} \\
&\quad + \mathscr{L} \left\{ x \in \mathbb{R}^n : |g(x)| > \gamma\lambda \right\}
\end{aligned}$$

by invoking $(\mathcal{I}_\lambda)$. Therefore, it follows that

$$\begin{aligned}
I_N &\le \kappa^p \varepsilon(\kappa,\gamma) \int_0^{\frac{N}{k}} \lambda^{p-1} \mathscr{L} \left\{ x \in \mathbb{R}^n : |f(x)| > \lambda \right\} \, d\lambda \\
&\quad + \kappa^p \int_0^{\frac{N}{k}} \lambda^{p-1} \mathscr{L} \left\{ x \in \mathbb{R}^n : |g(x)| > \gamma\lambda \right\} \, d\lambda \\
&\le \kappa^p \varepsilon(\kappa,\gamma) I_N + \left(\frac{\kappa}{\gamma}\right)^p \int_0^\infty \lambda^{p-1} \mathscr{L} \left\{ x \in \mathbb{R}^n : |g(x)| > \lambda \right\} \, d\lambda.
\end{aligned}$$

By the assumption that $\|f\|_{p_0} < \infty$, we have that $I_N < \infty$ when $p \in [p_0, \infty)$ and so

$$(1 - \kappa^p \varepsilon(\kappa,\gamma)) I_N \le \left(\frac{\kappa}{\gamma}\right)^p \int_0^\infty \lambda^{p-1} \mathscr{L} \left\{ x \in \mathbb{R}^n : |g(x)| > \lambda \right\} \, d\lambda.$$

Then, apply the monotone convergence theorem to obtain the conclusion. $\qquad\square$

The goal is to prove the following important inequality.

Theorem 5.3.8 (Fefferman-Stein inequality). *Let $p_0 \in (0, \infty)$ and $f \in \mathrm{L}^1_{\mathrm{loc}}(\mathbb{R}^n)$ such that $\|\mathcal{M}^{\square'} f\|_{p_0} < \infty$. Then, for all $p \in [p_0, \infty)$ there exists a $C_p > 0$ (independent of f) such that $\|\mathcal{M}^{\square'} f\|_p \le C_p \|\mathcal{M}^\sharp f\|_p$.*

Corollary 5.3.9. *Let $p \in (1, \infty)$. Then, there exists a $C(p, n) > 0$ such that for all $f \in \mathrm{L}^p(\mathbb{R}^n)$,*

$$\|\mathcal{M}^{\square'} f\|_p \le C(p, n) \|\mathcal{M}^\sharp f\|_p.$$

In particular, $\|f\|_p \simeq \|\mathcal{M}^{\square'} f\|_p \simeq \|\mathcal{M}^\sharp f\|_p$ on $\mathrm{L}^p(\mathbb{R}^n)$.

Proof. Apply the theorem with $p_0 = p$ since $f \in \mathrm{L}^p(\mathbb{R}^n)$ if and only if $\mathcal{M}^{\square'} f \in \mathrm{L}^p(\mathbb{R}^n)$. $\quad\square$

To prove the Fefferman-Stein inequality, by Proposition 5.3.7, it suffices to prove $(\mathcal{I}_\lambda)$ with f replaced with $\mathcal{M}^{\square'} f$ and g replaced with $\mathcal{M}^\sharp f$. First, we need two key Lemmas.

Lemma 5.3.10 (Localisation for maximal functions). *There exists $\kappa_0 = \kappa_0(n) > 1$ such that for all $f \in L^1_{loc}(\mathbb{R}^n)$, for all cubes Q, and all $\lambda > 0$ if there exists $C > 1$ and $\tilde{x} \in CQ$ with $\mathcal{M}^{\square'}f(\tilde{x}) \leq \lambda$, then for all $\kappa \geq \kappa_0$*

$$Q \cap \left\{ x \in \mathbb{R}^n : \mathcal{M}^{\square'}f(x) > \kappa\lambda \right\} \subset \left\{ x \in \mathbb{R}^n : \mathcal{M}^{\square'}(f\chi_{MQ})(x) > \frac{\kappa}{\kappa_0}\lambda \right\}$$

with $M = C + 2$.

Proof. We know that $\mathcal{M}^{\square'}f \leq \kappa_0 \mathcal{M}^{\square}f$ for some $\kappa_0 = \kappa_0(n) > 1$. Let

$$x \in Q \cap \left\{ x \in \mathbb{R}^n : \mathcal{M}^{\square'}f(x) > \kappa\lambda \right\}$$

and so

$$\mathcal{M}^{\square}f(x) > \frac{\kappa}{\kappa_0}\lambda \geq \lambda$$

since $\kappa \geq \kappa_0$. So, there exists an $r > 0$ such that

$$\frac{1}{|Q(x,r)|} \int_{Q(x,r)} |f(y)| \, d\mathcal{L}(y) > \frac{\kappa}{\kappa_0}\lambda.$$

First, $\tilde{x} \notin Q(x,r)$ since $\mathcal{M}^{\square'}f(x) \leq \lambda$. This implies that $\|x - \tilde{x}\|_\infty \geq r$. Secondly, by hypothesis, $\tilde{x} \in CQ$ and letting x_Q be the centre of Q,

$$\|x - \tilde{x}\|_\infty \leq \|x - x_Q\|_\infty + \|x_Q - \tilde{x}\|_\infty \leq \operatorname{rad} Q + C \operatorname{rad} Q \leq (C+1)\operatorname{rad} Q.$$

So, for any $y \in Q(x,r)$

$$\|y - x_Q\|_\infty \leq \|y - x\|_\infty + \|x - x_Q\|_\infty \leq r + \operatorname{rad} Q \leq (C+1)\operatorname{rad} Q + \operatorname{rad} Q \leq M \operatorname{rad} Q$$

Thus,

$$\mathcal{M}(f\chi_{MQ})(x) \geq \fint_{Q(x,r)} |f\chi_{MQ}| \, d\mathcal{L} = \fint_{Q(x,r)} |f| \, d\mathcal{L} > \frac{\kappa}{\kappa_0}\lambda$$

and completes the proof. $\qquad\square$

Lemma 5.3.11 (Proving $(\mathcal{I}_\lambda)$). *Fix $q \in (1, \infty]$ and $\alpha \geq 1$. Let $F \in L^1_{loc}(\mathbb{R}^n)$, $F \geq 0$ such that for all cubes Q there exists $G_Q, H_Q : \mathbb{R}^n \to \mathbb{R}_+$ measurable with*

(i) *$F \leq G_Q + H_Q$ almost everywhere $x \in Q$,*

(ii) *For all $x \in \mathbb{R}^n$,*

$$\alpha \mathcal{M}^{\square'}F(x) \geq \begin{cases} \sup_{Q \ni x} \left(\fint_Q H_Q^q \right)^{\frac{1}{q}} & q < \infty \\ \|H_Q\|_{L^\infty(Q)} & q = \infty \end{cases}.$$

Set

$$G(x) = \sup_{Q \ni x} \fint_Q G_Q.$$

Then, there exists $C = C(q,n) > 0$ and $\kappa_0' = \kappa_0'(\alpha, n) \geq 1$ such that for all $\lambda > 0$, $\kappa > \kappa_0'$, and $\gamma \in (0,1]$, $(\mathcal{I}_\lambda)$ holds for $\mathcal{M}^{\square'}F$ and G in place of f and g respectively. That is,

$$\mathscr{L}\left\{x \in \mathbb{R}^n : \left|\mathcal{M}^{\square'}F(x)\right| > \kappa\lambda, \ |G(x)| \leq \gamma\lambda\right\} \leq \varepsilon(\kappa, \gamma) \, \mathscr{L}\left\{x \in \mathbb{R}^n : \left|\mathcal{M}^{\square'}F(x)\right| > \lambda\right\}$$

where

$$\varepsilon(\kappa, \gamma) = C\left(\left(\frac{\alpha}{\kappa}\right)^q + \frac{\gamma}{\kappa}\right).$$

Remark 5.3.12. *When $q = \infty$, $\left(\frac{\alpha}{\kappa}\right)$ is replaced by 0.*

Proof. Let $\lambda > 0$ and $\mathcal{E}_\lambda = \left\{x \in \mathbb{R}^n : \mathcal{M}^{\square'}F(x) > \lambda\right\}$ is open by the lower semi-continuity of $\mathcal{M}^{\square'}F$.

If $\mathcal{E}_\lambda = \mathbb{R}^n$, there's nothing to do. So, suppose that $\mathcal{E}_\lambda \neq \mathbb{R}^n$ and use a Whitney decomposition with dyadic cubes (Theorem 2.3.1). So, $\mathcal{E}_\lambda = \sqcup \mathcal{Q}_i$, mutually disjoint with $\operatorname{diam} \mathcal{Q}_i$ comparable with $\operatorname{dist}(\mathcal{Q}_i, {}^c\mathcal{E}_\lambda)$. In particular, there exists a constant $C = C(n) > 1$ such that for all i, $C\mathcal{Q}_i \cap {}^c\mathcal{E}_\lambda \neq \varnothing$. That is, for each i, there exists a $\tilde{x}_i \in C\mathcal{Q}_i$ such that $\mathcal{M}^{\square'}f(\tilde{x}_i) \leq \lambda$. Set $D_i = \mathcal{Q}_i \cap \left\{x \in \mathbb{R}^n : \mathcal{M}^{\square'}f(x) > \kappa\lambda, \ G \leq \gamma\lambda\right\}$, and so

$$\sum_i \mathscr{L}(D_i) = \mathscr{L}\left\{x \in \mathbb{R}^n : \mathcal{M}^{\square'}f(x) > \kappa\lambda, \ G \leq \gamma\lambda\right\}$$

since $\kappa \geq 1$.

We estimate each D_i for each i. Assume $D_i \neq \varnothing$. So, there exists a $y_i \in \mathcal{Q}_i$ such that $G(y_i) \leq \gamma\lambda$. So, by the Localisation Lemma 5.3.10

$$\mathscr{L}(D_i) \leq \mathscr{L}\left(\mathcal{Q}_i \cap \left\{x \in \mathbb{R}^n : \mathcal{M}^{\square'}f(x) > \kappa\lambda\right\}\right)$$
$$\leq \mathscr{L}\left(\left\{x \in \mathbb{R}^n : \mathcal{M}(f\chi_{M\mathcal{Q}_i})(x) > \frac{\kappa}{\kappa_0}\lambda\right\}\right)$$
$$\leq A + B$$

where

$$A = \mathscr{L}\left\{x \in \mathbb{R}^n : \mathcal{M}^{\square'}(G_{M\mathcal{Q}_i}\chi_{M\mathcal{Q}_i})(x) > \frac{\kappa}{2\kappa_0}\lambda\right\}$$

and

$$B = \mathscr{L}\left\{x \in \mathbb{R}^n : \mathcal{M}^{\square'}(H_{M\mathcal{Q}_i}\chi_{M\mathcal{Q}_i})(x) > \frac{\kappa}{2\kappa_0}\lambda\right\}.$$

We estimate A by invoking the Maximal theorem (weak type $(1,1)$):

$$A \leq 2C\frac{\kappa_0}{\kappa\lambda}\int_{\mathbb{R}^n} G_{M\mathcal{Q}_i}\chi_{M\mathcal{Q}_i}\, d\mathscr{L} \leq 2C\frac{\kappa_0}{\kappa\lambda}\int_{M\mathcal{Q}_i} G_{M\mathcal{Q}_i}\, d\mathscr{L}$$
$$\leq 2C\frac{\kappa_0}{\kappa}\mathscr{L}(M\mathcal{Q}_i)\frac{G(y_i)}{\lambda} \leq 2C\frac{\kappa_0}{\kappa}\mathscr{L}(M\mathcal{Q}_i)\gamma.$$

By the Maximal theorem (weak type (q,q)) for $q < \infty$,

$$B \le C\left(\frac{2\kappa_0}{\kappa\lambda}\right)^q \int_{\mathbb{R}^n} (H_{M\mathcal{Q}_i}\chi_{M\mathcal{Q}_i})^q \, d\mathscr{L} = C\left(\frac{2\kappa_0}{\kappa\lambda}\right)^q \int_{M\mathcal{Q}_i} (H_{M\mathcal{Q}_i})^q \, d\mathscr{L}$$

$$\le C\left(\frac{2\kappa_0}{\kappa\lambda}\right)^q \mathscr{L}(M\mathcal{Q}_i)\left(\mathcal{M}^{\square'}f(\tilde{x}_i)\right)^q \le C\left(\frac{2\kappa_0}{\kappa\lambda}\right)^q \mathscr{L}(M\mathcal{Q}_i)(\alpha\lambda)^q = \mathscr{L}(M\mathcal{Q}_i)C\left(\frac{2\kappa_0}{\kappa}\right)^q \mathscr{L}(M\mathcal{Q}_i)\alpha^q.$$

If $q = \infty$, then

$$\|H_{M\mathcal{Q}_i}\|_{\mathrm{L}^\infty(M\mathcal{Q}_i)} \le \alpha\mathcal{M}^{\square'}F(\tilde{x}_i) \le \alpha\lambda.$$

Thus, if $\kappa > \alpha 2\kappa_0$, then

$$\left\{x \in \mathbb{R}^n : \mathcal{M}^{\square'}(H_{M\mathcal{Q}_i}\chi_{M\mathcal{Q}_i})(x) > \frac{\kappa}{\kappa_0}\lambda\right\} = \varnothing.$$

$\square$

We are now in a position to prove the Fefferman-Stein inequality.

Proof of the Fefferman-Stein inequality. By hypothesis, $f \in \mathrm{L}^1_{\mathrm{loc}}(\mathbb{R}^n)$ such that $\|\mathcal{M}^{\square'}f\|_{p_0} < \infty$. Set $F = |f|$. Pick a cube Q and let $G_Q = |f - \mathrm{m}_Q\, f|$ and $H_Q = |\mathrm{m}_Q\, f|$. Then,

(i) $F \le G_Q + H_Q$, and

(ii) $\|H_Q\|_{\mathrm{L}^\infty(Q)} = |\mathrm{m}_Q\, f| \le \mathcal{M}^{\square'}f(x) = \mathcal{M}'F(x)$ for all $x \in Q$.

Apply Lemma 5.3.11 to get the inequality $(\mathcal{I}_\lambda)$ with $G = \mathcal{M}^\sharp f$ and $\varepsilon(\kappa,\gamma) = C\frac{\gamma}{\kappa}$.

Then, apply Proposition 5.3.7 with $p \ge p_0$ since $1 - \kappa^p\varepsilon(\kappa,\gamma) > 0$ for fixed κ and small γ. Thus, we conclude that for some $C_p > 0$,

$$\|\mathcal{M}^{\square'}f\|_p = \|\mathcal{M}'F\|_p \le C_p\|G\|_p = C_p\|\mathcal{M}^\sharp f\|_p$$

and the proof is complete.

$\square$

We have the following Corollary to the Fefferman-Stein inequality.

Corollary 5.3.13 (Stampacchia). *Suppose that T is sublinear on $\mathcal{D}_{\mathbb{R}^n}$, a subspace of the space of measurable functions stable under multiplication by indicator functions. Suppose further that $T : \mathrm{L}^p(\mathbb{R}^n) \to \mathrm{L}^p(\mathbb{R}^n)$ for some $p \in [1,\infty)$ and $T : \mathcal{D}_{\mathbb{R}^n} \cap \mathrm{L}^\infty(\mathbb{R}^n) \to \mathrm{BMO}$ are bounded. Then for all $q \in (p,\infty)$, T is strong type (q,q) with log convex control of operators "norms."*

f

52

Chapter 6

Hardy Spaces

6.1 Atoms and H^1

Hardy spaces are function spaces designed to be better suited to some applications than L^1. We consider *atomic* Hardy spaces.

Definition 6.1.1 (∞-atom). *Let Q be a cube in $\mathbb{R}^n$. A measurable function $a : Q \to \mathbb{C}$ is called an ∞-atom on Q if*

    (i) spt $a \subset Q$,

    (ii) $\|a\|_\infty \leq \frac{1}{\mathscr{L}(Q)}$,

    (iii) $\int_Q a \, d\mathscr{L} = 0$.

We denote the collection of ∞-atoms on Q by $\mathscr{A}_Q^\infty$ and $\mathscr{A}^\infty = \cup_Q \mathscr{A}_Q^\infty$.

Remark 6.1.2. *Note that (i) along with (ii) implies that $\|a\|_1 \leq 1$.*

Definition 6.1.3 (p-atom). *Let Q be a cube in $\mathbb{R}^n$. A measurable function $a : Q \to \mathbb{C}$ is called an p-atom on Q if*

    (i) spt $a \subset Q$,

    (ii) $\|a\|_p \leq \frac{1}{\mathscr{L}(Q)^{1-\frac{1}{p}}}$,

    (iii) $\int_Q a \, d\mathscr{L} = 0$. *We denote the collection of p-atoms on Q by $\mathscr{A}_Q^p$ and $\mathscr{A}^p = \cup_Q \mathscr{A}_Q^p$.*

Definition 6.1.4 ($H^{1,p}$). *Let $1 < p \leq \infty$ and $f \in L^1(\mathbb{R}^n)$. We say that $f \in H^{1,p}$ if there exist p-atoms $\{a_i\}_{i\in\mathbb{N}}$ and $(\lambda_i)_{i\in\mathbb{N}} \in \ell^1(\mathbb{N})$ such that*

$$f = \sum_{i=0}^{\infty} \lambda_i a_i$$

almost everywhere.

Remark 6.1.5. *The fact that $\sum |\lambda_i| < \infty$ combined with $\|a_i\|_1 \le 1$ implies that $\sum \lambda_i a_i$ converges in $L^1(\mathbb{R}^n)$. For $f \in L^1(\mathbb{R}^n)$, then the equality $f = \sum \lambda_i a_i$ is in $L^1(\mathbb{R}^n)$. We could certainly give a similar definition in the more general setting of $f \in \mathscr{S}'(\mathbb{R}^n)$. Then, we would ask the convergence of the series in the sense of $\mathscr{S}'(\mathbb{R}^n)$ in the definition. Then as $L^1(\mathbb{R}^n)$ embeds in $\mathscr{S}'(\mathbb{R}^n)$ it coincides with the L^1 function $\sum_i \lambda_i a_i$ after identification.*

On noting that $H^{1,p}$ is a $\mathbb{K}$ vector space, we define a norm.

Definition 6.1.6 ($H^{1,p}$ norm). *We define*

$$\|f\|_{H^{1,p}} = \inf \left\{ \sum |\lambda_i| : f = \sum \lambda_i a_i \right\},$$

where the infimum is taken over all possible representations of f.

Proposition 6.1.7. *(i) $(H^{1,p}, \|\cdot\|_{H^{1,p}})$ is a Banach space,*

(ii) Whenever $1 < p < r < \infty$, we have

$$H^{1,\infty} \subset H^{1,r} \subset H^{1,p} \subset L^1(\mathbb{R}^n),$$

Proof. Exercise. $\qquad\square$

Theorem 6.1.8 (Equivalence of $H^{1,p}$ spaces). *For $1 < p < \infty$, $H^{1,\infty} = H^{1,p}$ with equivalence of norms.*

Proof. (i) We establish what is called the *Calderón-Zygmund decomposition* of functions. More precisely, we show given a p-atom a that there exists a decomposition $a = b + g$ with $b \in H^{1,p}$ with $\|b\|_{H^{1,p}} \le \frac{1}{2}$ and $g \in H^{1,\infty}$ with $\|g\|_{H^{1,\infty}} \le C(n,p)$ (Note that $\|a\|_{H^{1,p}} \le 1$ so $\|b\|_{H^{1,p}} \le \frac{1}{2}$ is better).

Let Q be a cube in $\mathbb{R}^n$ such that $a \in \mathscr{A}_Q^p$. As before, let $\mathscr{D}(Q)$ denote the dyadic subcubes of Q. We have

$$m_Q |a|^p = \frac{1}{\mathscr{L}(Q)} \int_Q |a|^p \le \frac{1}{\mathscr{L}(Q)^p}.$$

Fix $\alpha > 0$ with $\alpha^p > m_Q |a|^p$ to be chosen later and let

$$\mathcal{E}_\alpha = \left\{ x \in Q : (\mathcal{M}^{\mathscr{Q}} |a|^p)(x)^{\frac{1}{p}} > \alpha \right\}.$$

Note then that $Q \ne \mathcal{E}_\alpha$ and $\mathcal{E}_\alpha = \sqcup_{i=1}^\infty \mathcal{Q}_i$ where $\mathcal{Q}_i \in \mathscr{D}(Q)$ maximal for the property that $m_{\mathcal{Q}_i} |a|^p > \alpha^p$. Set

$$b_i = (a - m_{\mathcal{Q}_i} a) \chi_{\mathcal{Q}_i}$$

and $b = \sum_{i=1}^\infty b_i$. Let $g = a - b$.

We note the properties of b_i. First, $\operatorname{spt} b_i \subset \overline{\mathcal{Q}_i}$ and

$$\int_{\mathcal{Q}_i} b_i \, d\mathscr{L} = 0.$$

Also,

$$\|b_i\|_p \leq \left(\int_{\mathcal{Q}_i} |a|^p\right)^{\frac{1}{p}} + |\mathrm{m}_{\mathcal{Q}_i}\, a|\, \mathscr{L}(\mathcal{Q}_i)^{\frac{1}{p}} = \frac{1}{\mathscr{L}(\mathcal{Q}_i)^{1-\frac{1}{p}}} \lambda_i$$

where $\lambda_i = 2\,|\mathrm{m}_{\mathcal{Q}_i}\, |a|^p|^{\frac{1}{p}}\, \mathscr{L}(\mathcal{Q}_i)$. We note that $\lambda_i \neq 0$. For otherwise, if $\lambda_i = 0$, then $a = 0$ on $\mathcal{Q}_i$ but we assume that $(\mathrm{m}_{\mathcal{Q}_i}\, |a|^p)^{\frac{1}{p}} > \alpha > 0$ which is a contradiction. Therefore, $a_i = \frac{1}{\lambda_i} b_i$ is a p-atom.

Now, by Hölder's inequality and the Maximal Theorem,

$$\sum_{i=1}^{\infty} \lambda_i = 2\sum_{i=1}^{\infty} \left|\int_{\mathcal{Q}_i} |a|^p\right|^{\frac{1}{p}} \mathscr{L}(\mathcal{Q}_i)^{1-\frac{1}{p}} \leq 2\left(\sum_{i=1}^{\infty} \int_{\mathcal{Q}_i} |a|^p\right)^{\frac{1}{p}} \left(\sum_{i=1}^{\infty} \mathscr{L}(\mathcal{Q}_i)\right)^{1-\frac{1}{p}}$$

$$\leq 2\left(\int_Q |a|^p\right)^{\frac{1}{p}} \mathscr{L}(\mathcal{E}_\alpha)^{1-\frac{1}{p}} \leq 2\left(\int_Q |a|^p\right) \frac{C}{\alpha^{p-1}} \leq 2\left(\frac{1}{\alpha \mathscr{L}(Q)}\right)^{p-1}.$$

Now, we choose α such that

$$2\left(\frac{1}{\alpha \mathscr{L}(Q)}\right)^{p-1} = \frac{1}{2}$$

to find

$$\alpha = \frac{c_p}{\mathscr{L}(Q)},$$

with $c_p = 4^{\frac{1}{p-1}}$. It then follows that $\|b\|_{\mathrm{H}^{1,p}} \leq \frac{1}{2}$.

Now, consider g,

$$g = \begin{cases} a & \text{on } Q \setminus \mathcal{E}_\alpha \\ \mathrm{m}_{\mathcal{Q}_i}\, a & \text{on } \mathcal{Q}_i \text{ for each } i \end{cases}.$$

Certainly, on $Q \setminus \mathcal{E}_\alpha$, $|a|^p \leq \mathcal{M}^{\mathscr{Q}}(|a|^p) \leq \alpha^p$ almost everywhere and so $|g| \leq \alpha$ almost everywhere. On $\mathcal{Q}_i$, by maximality and Hölder's inequality,

$$|\mathrm{m}_{\mathcal{Q}_i}\, a| \leq \mathrm{m}_{\mathcal{Q}_i} |a| \leq 2^n\, \mathrm{m}_{\widehat{\mathcal{Q}}_i} |a| \leq 2^n \alpha.$$

Hence, $|g| \leq 2^n \alpha$. It then follows that

$$\|g\|_\infty \leq 2^n \alpha = 2^n \frac{c_p}{\mathscr{L}(Q)}.$$

We also have $\int_Q g = \int_Q a = 0$, so $\frac{1}{2^n c_p} g \in \mathscr{A}_Q^\infty$ which implies that $g \in \mathrm{H}^{1,\infty}$ with $\|g\|_{\mathrm{H}^{1,\infty}} \leq 2^n c_p$.

(ii) Fix $f_0 \in \mathrm{H}^{1,p}$ with $f_0 \neq 0$. We show there exists a decomposition $f_0 = f_1 + g^0$ with

$$\|f_1\|_{\mathrm{H}^{1,p}} \leq \frac{2}{3}\|f_0\|_{\mathrm{H}^{1,p}} \qquad \text{and} \qquad \|g^0\|_{\mathrm{H}^{1,\infty}} \leq \frac{4}{3} 2^n c_p \|f_0\|_{\mathrm{H}^{1,p}}.$$

For every $\varepsilon > 0$ there exists an atomic decomposition $f_0 = \sum_{i=1}^\infty \lambda_j a_j$ with

$$\sum_{i=1}^\infty |\lambda_j| \leq \|f_0\|_{\mathrm{H}^{1,p}} + \varepsilon.$$

55

We apply (i) to each a_j to find a decomposition $a_j = b_j + g_j$ with $\|b_j\|_{\mathrm{H}^{1,p}} \leq \frac{1}{2}$. Certainly, $f_1 = \sum_{j=1}^{\infty} \lambda_j b_j$ exists since $\mathrm{H}^{1,p}$ is a Banach space (Proposition 6.1.7), and

$$\|f_1\|_{\mathrm{H}^{1,p}} \leq \frac{1}{2} \sum_{j=1}^{\infty} |\lambda_j| \leq \frac{1}{2} \left(\|f_0\|_{\mathrm{H}^{1,p}} + \varepsilon \right).$$

So, choose $\varepsilon = \frac{1}{3} \|f_0\|_{\mathrm{H}^{1,p}}$. Setting $g^0 = \sum_{j=0}^{\infty} \lambda_j g_j$ (the sum converging in $\mathrm{H}^{1,\infty}$ because it is a Banach space and $\|g_j\|_{\mathrm{H}^{1,\infty}} \leq 2^n c_p$), we find

$$\|g^0\|_{\mathrm{H}^{1,\infty}} \leq 2^n c_p \left(\|f_0\|_{\mathrm{H}^{1,p}} + \varepsilon \right) = \frac{4}{3} 2^n c_p \|f_0\|_{\mathrm{H}^{1,p}}.$$

(iii) We iterate

$$f_0 = f_1 + g^0$$
$$f_1 = f_2 + g^1$$
$$f_2 = f_3 + g^2$$
$$\vdots = \quad \vdots$$

and so for each k,

$$f_0 = f_k + g^0 + g^1 + \cdots + g^{k-1}.$$

Certainly $f_k \to 0$ in $\mathrm{H}^{1,p}$, and $g^0 + g^1 + \cdots + g^k$ converges to g in $\mathrm{H}^{1,\infty}$ with

$$\|g\|_{\mathrm{H}^{1,\infty}} \leq \frac{4}{3} 2^n c_p \sum_{j=0}^{\infty} \left(\frac{2}{3} \right)^j \|f_0\|_{\mathrm{H}^{1,p}}.$$

By Proposition 6.1.7, convergence in $\mathrm{H}^{1,\infty}$ implies convergence in $\mathrm{H}^{1,p}$ and so $f_0 = g$ and $f_0 \in \mathrm{H}^{1,\infty}$.

$\square$

This motivates the following definition.

Definition 6.1.9 (Hardy space H^1). *We define H^1 to be any $\mathrm{H}^{1,p}$ for $1 < p \leq \infty$ with the corresponding norm.*

6.2 $\mathrm{H}^1 - \mathrm{BMO}$ **Duality**

We show that the dual space of H^1 and BMO are isomorphic with equivalent norms. This relationship was first established by C. Fefferman but using a different characterisation of H^1.

Theorem 6.2.1 ($\mathrm{H}^1 - \mathrm{BMO}$ duality). *The dual of H^1 is isomorphic to BMO with equivalent norms.*

Not quite a correct "proof". We work with $H^1 = H^{1,2}$ and $BMO = BMO_2$ with corresponding norms $\|\cdot\|_{H^{1,2}}$ and $\|\cdot\|_*$. Let $b \in BMO_2$ and $a \in \mathscr{A}^2$. Set

$$L_b(a) = \int_{\mathbb{R}^n} b(x)a(x) \, d\mathscr{L}(x).$$

Then, $L_b(a)$ is well defined and since for all cubes Q we have spt $a \subset Q$ and $\int_Q a = 0$,

$$|L_b(a)| \leq \left(\int_Q |b - m_Q b| \right)^{\frac{1}{2}} \|a\|_2 \leq \|b\|_*.$$

By linearity, L_b is defined on Vect $\mathscr{A}^2$. Let $f \in$ Vect $\mathscr{A}^2$. Then, $f = \sum_{i=1}^m \lambda_i a_i$ and

$$L_b(f) \leq \|b\|_* \sum_{i=1}^m \lambda_i a_i.$$

By density of Vect $\mathscr{A}^2$ in $H^{1,2}$, L_b can be defined on all of $H^{1,2}$. $\qquad\square$

Remark 6.2.2. *This proof is correct if we know that for all $f \in$ Vect $\mathscr{A}^2$, $\|f\|_{H^{1,2}} \simeq \inf \sum_{finite} |\lambda_i|$. Note that we automatically have $\|f\|_{H^{1,2}} \leq \inf \sum_{finite} |\lambda_i|$. This subtlety went unnoticed for some time, and it has only been recently that $\|f\|_{H^{1,2}} \gtrsim \inf \sum_{finite} |\lambda_i|$ was proved by an intricate argument. Also this equivalence is not true for all atoms - there exists a counter example for ∞-atoms.*

We take a different approach in proving the theorem.

Proof of the $H^1 - BMO$ duality. We work with $H^1 = H^{1,2}$ and $BMO = BMO_2$ with corresponding norms $\|\cdot\|_{H^{1,2}}$ and $\|\cdot\|_*$.

(i) Take $b \in L^\infty(\mathbb{R}^n)$ and $f \in H^{1,2}$. Define,

$$L_b(a) = \int_{\mathbb{R}^n} b(x)a(x) \, d\mathscr{L}(x)$$

and note that it is well defined since $H^{1,2} \subset L^1(\mathbb{R}^n)$. If $f = \sum_{i=1}^\infty \lambda_i a_i$, we can apply Dominated Convergence since

$$\int_{\mathbb{R}^n} \sum_{i=1}^\infty |b(x)\lambda_i a_i(x)| \, d\mathscr{L}(x) \leq \|b\|_\infty \sum_{i=1}^\infty |\lambda_i|.$$

Also,

$$|L_b(a)| = \left| \int_{\mathbb{R}^n} (b(x) - m_Q b) \, a_i(x) \, d\mathscr{L}(x) \right| \leq \|b\|_\infty$$

if spt $a_i \subset Q$. This implies that $|L_b(f)| \leq \|b\|_* \sum_{i=1}^\infty |\lambda_i|$ and taking an infimum over all possible λ_i,

$$|L_b(f)| \leq \|b\|_* \|f\|_{H^{1,2}}.$$

(ii) Now take $b \in \mathrm{BMO}_2$ and $f \in \operatorname{Vect} \mathscr{A}^2$. Let $(b_k)_{k=1}^{\infty}$ be the truncation of b. So, $|b_k| \leq |b|$ almost everywhere, $b_k \nearrow b$ almost everywhere, and $\|b_k\|_* \leq 2\|b\|_*$. [1] Now, suppose $f = \sum_{i=1}^m \lambda_i a_i$ and $L_{b_k}(f) = \sum_{i=1}^m \lambda_i L_{b_k}(a_i)$. Since $b \in \mathrm{BMO}_2$, we have that $b \in \mathrm{L}^2_{\mathrm{loc}}(\mathbb{R}^n)$ which implies $b \in \mathrm{L}^2(\operatorname{spt} a_i)$. So, certainly $|b_k a_i| \leq |b a_i| \in \mathrm{L}^1(\mathbb{R}^n)$ almost everywhere. Thus,

$$\int_{\mathbb{R}^n} b_k a_i \, d\mathscr{L} \to \int_{\mathbb{R}^n} b a_i \, d\mathscr{L},$$

ie., $L_{b_k}(a_i) \to L_b(a)$. This implies that

$$|L_b(f)| \leq \sup_k |L_{b_k}(f)| \leq 2\|b\|_* \|f\|_{\mathrm{H}^{1,2}}.$$

(iii) We now apply a density argument and extend L_b to the whole of $\mathrm{H}^{1,2}$. Let $\tilde{L}_b$ denote this extension.

So we have shown that whenever $b \in \mathrm{BMO}_2$ we have $\tilde{L}_b \in (\mathrm{H}^{1,2})'$. Let $T : \mathrm{BMO}_2 \to (\mathrm{H}^{1,2})'$ denote the map $b \mapsto \tilde{L}_b$.

(iv) We leave it as an exercise to show that T is linear.

(v) We show that T is injective. Let $b \in \mathrm{BMO}_2$ such that $\tilde{L}_b = 0$. We show that b is constant.

Fix a cube Q and let $f \in \mathrm{L}^2(Q)$ with $\int_Q f \, d\mathscr{L} = 0$. Then, there exists a $\lambda \in \mathbb{K}$ such that $\frac{f}{\lambda} \in \mathscr{A}_Q^2$ which implies that $f \in \operatorname{Vect} \mathscr{A}^2$. So,

$$0 = \tilde{L}_b(f) = L_b(f) = \int_Q bf \, d\mathscr{L}$$

and since we assume $\int_Q f = 0$, it follows that $b|_Q$ is constant. By exhaustion of $\mathbb{R}^n$ by increasing Q, we deduce that b is constant.

(vi) Lastly, we show that T is surjective. Let $L \in (\mathrm{H}^{1,2})'$ and fix a cube Q. Let $\mathrm{L}_0^2(Q) = \left\{ f \in \mathrm{L}^2(Q) : \int_Q f \, d\mathscr{L} = 0 \right\}$ and note that $\mathrm{L}_0^2(Q) \subset \mathrm{H}^{1,2}$. Take $f \in \mathrm{L}_0^2(Q)$ and $\lambda = \|f\|_2 \mathscr{L}(Q)^{\frac{1}{2}}$. Then,

$$\left\| \frac{f}{\lambda} \right\|_2 \leq \frac{1}{\mathscr{L}(Q)^{\frac{1}{2}}}$$

so $\frac{f}{\lambda}$ is a 2-atom. Thus,

$$\|f\|_{\mathrm{H}^{1,2}} = \lambda \left\| \frac{f}{\lambda} \right\|_{\mathrm{H}^{1,2}} \leq \|f\|_2 \mathscr{L}(Q)^{\frac{1}{2}}.$$

Furthermore, $L|_{\mathrm{L}_0^2(Q)} \in (\mathrm{L}_0^2(Q))'$ and so by the Riesz Representation Theorem, there exists a $b_Q \in \mathrm{L}_0^2(Q)$ such that for all $f \in \mathrm{L}_0^2(Q)$,

$$L(f) = \int_Q b_Q f \, d\mathscr{L}$$

and $\|b_Q\|_2 \leq \|L\|_{(\mathrm{H}^{1,2})'} \mathscr{L}(Q)^{\frac{1}{2}}$.

---

[1] Here we assume b real valued. For complex valued b, separate real and imaginary parts.

Let Q, Q' denote two cubes with $Q \subset Q'$. Then, whenever $f \in \mathrm{L}_0^2(Q)$,

$$L(f) = \int_Q b_Q f \, d\mathscr{L} = \int_{Q'} b_{Q'} f \, d\mathscr{L}$$

and so $b_Q - b_{Q'}$ is constant almost everywhere in Q. Define b as follows:

$$b(x) = \begin{cases} b_{[-1,1]^n}(x) & x \in [-1,1]^n \\ b_{[-2^j,2^j]^n}(x) + c_j & x \in [-2^j, 2^j]^n \setminus [-2^{j-1}, 2^{j-1}]^n, \ j \geq 1 \end{cases}$$

where c_j is the constant such that $b_{[-2^j,2^j]^n} - b_{[-1,1]^n} = -c_j$ on $[-1,1]^n$.

We show that $b \in \mathrm{BMO}_2$, $\|b\|_* \leq \|L\|_{(\mathrm{H}^{1,2})'}$ and $L = \tilde{L}_b$. Fix Q, and let $j \in \mathbb{N}$ such that $Q \subset [-2^j, 2^j]^n$. Let k be such that $2 \leq k \leq j$. Then, $c_k - c_{k-1} = b_{[-2^k,2^k]^n} - b_{[-2^{k-1},2^{k-1}]^n}$ which is constant on $[-2^{k-1}, 2^{k-1}]^n$ and in particular on $[-2^{k-1}, 2^{k-1}]^n \setminus [-2^{k-2}, 2^{k-2}]^n$. Therefore, $b(x) = b_{[-2^j,2^j]^n}(x) + c_j$ on all of $[-2^j, 2^j]^n$ and in particular on Q. Also, there exists a constant c such that $b_{[-2^j,2^j]^n} - b_Q = c$ on the cube Q and so $b = b_Q + c + c_j$ on Q. Then,

$$b - \mathrm{m}_Q \, b = b_Q + c + c_j - \mathrm{m}_Q \, b_Q - c - c_j = b_Q$$

since $\mathrm{m}_Q \, b_Q = 0$. Therefore,

$$\int_Q |b - \mathrm{m}_Q \, b|^2 \, d\mathscr{L} = \int_Q |b_Q|^2 \, d\mathscr{L} \leq \|L\|_{(\mathrm{H}^{1,2})'}^2 \mathscr{L}(Q).$$

The fact that $L = \tilde{L}_b$ follows from the fact that $L(a) = \tilde{L}_b(a)$ for all $a \in \mathscr{A}^2$.

$\square$

Remark 6.2.3. BMO *and "atomic* H^1*" can be defined on any space of homogeneous type. The results of this section go through in such generality. See [CR77].*

Chapter 7

Calderón-Zygmund Operators

7.1 Calderón-Zygmund Kernels and Operators

We denote the diagonal of $\mathbb{R}^n \times \mathbb{R}^n$ by $\Delta = \{(x,x) : x \in \mathbb{R}^n\}$.

Definition 7.1.1 (Calderón-Zygmund Kernel). *Let $0 < \alpha \leq 1$. A Calderón-Zygmund Kernel of order α is a continuous function $K : {}^c\Delta \to \mathbb{K}$ such that there exist a $C > 0$ and satisfies:*

(i) *For all $(x,y) \in {}^c\Delta$,*

$$|K(x,y)| \leq \frac{C}{|x-y|^n},$$

(ii) *For all $x, y, y' \in \mathbb{R}^n$ satisfying $|y-y'| \leq \frac{1}{2}|x-y|$ when $x \neq y$,*

$$\left|K(x,y) - K(x,y')\right| \leq C \left(\frac{|y-y'|}{|x-y|}\right)^\alpha \frac{1}{|x-y|^n},$$

(iii) *For all $x, x', y \in \mathbb{R}^n$ satisfying $|x-x'| \leq \frac{1}{2}|x-y|$ when $x \neq y$,*

$$\left|K(x,y) - K(x',y)\right| \leq C \left(\frac{|x-x'|}{|x-y|}\right)^\alpha \frac{1}{|x-y|^n},$$

We write $K \in \mathrm{CZK}_\alpha$ and norm it via $\|K\|_\alpha = \inf\{C : \text{(i) to (iii) hold}\}$.

Remark 7.1.2. (i) *The constant $\frac{1}{2}$ can be replaced by any $\theta \in (0,1)$. Then, the constant C changes.*

(ii) *The Euclidean norm $|\cdot|$ can be changed to any other norm. Again, C changes.*

(iii) *When $\alpha = 1$, $\nabla_y K(x,y)$ exists almost everywhere and satisfies:*

$$|\nabla_y K(x,y)| \leq \frac{C'}{|x-y|^{n+1}}$$

for all $(x,y) \in {}^c\Delta$.

(iv) When $\alpha = 1$, $\nabla_x K(x,y)$ exists almost everywhere and satisfies:

$$|\nabla_x K(x,y)| \leq \frac{C'}{|x-y|^{n+1}}$$

for all $(x,y) \in {}^c\Delta$.

(v) Define $K^*(x,y) = \overline{K(y,x)}$. Then, $K \in \mathrm{CZK}_\alpha$ implies $K^* \in \mathrm{CZK}_\alpha$.

Definition 7.1.3 (Kernel associated to an operator). *Let $T \in \mathcal{L}(\mathrm{L}^2(\mathbb{R}^n))$. We say that a kernel $K : {}^c\Delta \to \mathbb{K}$ is associated to T if for all $f \in \mathrm{L}^2(\mathbb{R}^n)$, with spt f compact,*

$$Tf(x) = \int_{\mathbb{R}^n} K(x,y)f(y) \, d\mathscr{L}(y)$$

for almost every $x \in {}^c(\mathrm{spt}\ f)$.

Remark 7.1.4. *This integral is a Lebesgue integral for all $x \in {}^c(\mathrm{spt}\ f)$. Moreover, this says that Tf can be represented by this integral away from the support of f.*

Definition 7.1.5 (Calderón-Zygmund Operator). *A Calderón-Zygmund Operator of order α is an operator $T \in \mathcal{L}(\mathrm{L}^2(\mathbb{R}^n))$ that is associated to a $K \in \mathrm{CZK}_\alpha$. We define CZO_α to be the collection of all Calderón-Zygmund operators of order α. Also, $\|T\|_{\mathrm{CZO}_\alpha} = \|T\|_{\mathcal{L}(\mathrm{L}^2(\mathbb{R}^n))} + \|K\|_\alpha$.*

Remark 7.1.6. *(i) $T \in \mathrm{CZO}_\alpha$ if and only if $T^* \in \mathrm{CZO}_\alpha$.*

Also, let $f,g \in \mathrm{L}^2(\mathbb{R}^n)$ with spt f, spt g compact and spt $f \cap$ spt $g = \varnothing$. Then,

$$\langle T^*g, f\rangle = \langle g, Tf\rangle = \int_{\mathbb{R}^n} g(x)\overline{\left(\int_{\mathbb{R}^n} K(x,y)f(y) \, d\mathscr{L}(y)\right)} \, d\mathscr{L}(x)$$

$$= \int_{\mathbb{R}^n} \overline{f(y)} \left(\int_{\mathbb{R}^n} \overline{K(x,y)}g(x) \, d\mathscr{L}(x)\right) \, d\mathscr{L}(y)$$

and since f was arbitrary, $T^*g(y) = \int_{\mathbb{R}^n} \overline{K(x,y)}g(x) \, d\mathscr{L}(x)$ for almost every $y \in {}^c(\mathrm{spt}\ g)$. That is, T^* has associated kernel K^*.

(ii) $T \in \mathrm{CZO}_\alpha$ if and only if $T^{\mathrm{tr}} \in \mathrm{CZO}_\alpha$, where T^{tr} is the real transpose of T. The associated kernel to T^{tr} is $K^{\mathrm{tr}}(x,y) = K(y,x)$.

(iii) The map $T \mapsto K$, where $T \in \mathrm{CZO}_\alpha$ and $K \in \mathrm{CZK}_\alpha$ the associated kernel, is not injective. Consequently, one cannot define a CZO_α uniquely given a kernel $K \in \mathrm{CZO}_\alpha$. The following is an important illustration.

Let $m \in \mathrm{L}^\infty(\mathbb{R}^n)$ and let T_m be the map $f \mapsto mf$. It is easy to check that this is a bounded operator on $\mathrm{L}^2(\mathbb{R}^n)$. Let $K = 0$ on ${}^c\Delta$ and let $f \in \mathrm{L}^2(\mathbb{R}^n)$ with spt f compact. Then, whenever $x \notin \mathrm{spt}\ f$, $T_m f(x) = m(x)f(x) = 0$. Therefore,

$$T_m f(x) = \int_{\mathbb{R}^n} K(x,y)f(y) \, d\mathscr{L}(y)$$

whenever $x \in {}^c(\mathrm{spt}\ f)$, which shows the associated kernel to T_m is 0.

7.2 The Hilbert Transform, Riesz Transforms, and The Cauchy Operator

We discuss three important examples that have motivated the theory.

7.2.1 The Hilbert Transform

Definition 7.2.1 (Hilbert Transform). *We define a map* $H : \mathscr{S}(\mathbb{R}^n) \to \mathscr{S}'(\mathbb{R}^n)$ *by*

$$H(\varphi) = \mathrm{p.\,v.} \left(\frac{1}{\pi x} \right) * \varphi.$$

That is,

$$\left(\mathrm{p.\,v.} \left(\frac{1}{\pi x} \right) * \varphi \right)(y) = \lim_{\varepsilon \to 0} \int_{\{x : |x| > \varepsilon\}} \frac{1}{\pi x} \varphi(y - x) \, d\mathscr{L}(x).$$

Proposition 7.2.2. *H extends to a bounded operator on* $\mathrm{L}^2(\mathbb{R})$.

Proof. We can analyse this convolution via the *Fourier Transform*. For a function $\varphi \in \mathscr{S}(\mathbb{R}^n)$, the Fourier transform is given by

$$\hat{\varphi}(\xi) = \int_{\mathbb{R}^n} \mathrm{e}^{-\imath x \cdot \xi} \, \varphi(x) \, d\mathscr{L}(x).$$

We can extend this naturally to $T \in \mathscr{S}'(\mathbb{R}^n)$ by defining $\hat{T}$ via $\langle \hat{T}, \varphi \rangle = \langle T, \hat{\varphi} \rangle$ for every $\varphi \in \mathscr{S}(\mathbb{R}^n)$. So, when $\varphi \in \mathscr{S}(\mathbb{R})$,

$$\langle \mathrm{p.\,v.} \left(\frac{1}{\pi x} \right)\hat{\,}, \varphi \rangle = \langle \mathrm{p.\,v.} \frac{1}{\pi x}, \hat{\varphi} \rangle$$

$$= \lim_{\varepsilon \to 0} \int_{\{x : |x| > \varepsilon\}} \frac{1}{\pi x} \hat{\varphi}(x) \, d\mathscr{L}(x)$$

$$= \lim_{\varepsilon \to 0} \int_{\{x : \varepsilon^{-1} > |x| > \varepsilon\}} \frac{1}{\pi x} \hat{\varphi}(x) \, d\mathscr{L}(x)$$

$$= \lim_{\varepsilon \to 0} \int_{\mathbb{R}^n} \varphi(\xi) \left(\int_{\{x : \varepsilon^{-1} > |x| > \varepsilon\}} \frac{1}{\pi x} \mathrm{e}^{-\imath x \cdot \xi} \, d\mathscr{L}(x) \right) d\mathscr{L}(\xi).$$

Now, fix $\xi \in \mathbb{R}$, $\xi \neq 0$. Then,

$$\int_{\{x : \varepsilon^{-1} > |x| > \varepsilon\}} \frac{1}{\pi x} \mathrm{e}^{-\imath x \cdot \xi} \, d\mathscr{L}(x) = -\imath \int_{\{x : \varepsilon^{-1} > |x| > \varepsilon\}} \frac{1}{\pi x} \sin(x \cdot \xi) \, d\mathscr{L}(x)$$

$$= -2\imath \int_{\{x : \varepsilon^{-1} > x > \varepsilon\}} \frac{1}{\pi x} \sin(x \cdot \xi) \, d\mathscr{L}(x)$$

$$= -2\imath \int_{\{x : \varepsilon^{-1} > x > \varepsilon\}} \frac{1}{\pi x} \sin(x \, |\xi|) \, \mathrm{sgn}(\xi) \, d\mathscr{L}(x)$$

$$= -\frac{2\imath}{\pi} \mathrm{sgn}(\xi) \int_{\frac{\varepsilon}{|\xi|}}^{\frac{1}{\varepsilon |\xi|}} \frac{\sin u}{u} \, d\mathscr{L}(u).$$

The integral appearing on the right hand side is uniformly bounded on ε and ξ and converges to $\frac{\pi}{2}$ when $\varepsilon \to 0$. Thus, by Dominated Convergence,

$$\langle \mathrm{p.\,v.} \left(\frac{1}{\pi x} \right), \varphi \rangle = \int_{\mathbb{R}} -\imath \, \mathrm{sgn}(\xi) \varphi(\xi) \, d\mathscr{L}(\xi)$$

and so for all $\varphi \in \mathscr{S}(\mathbb{R})$, $\widehat{H\varphi}(\xi) = -\imath \, \mathrm{sgn}(\xi)\hat{\varphi}(\xi)$. Since the Fourier transform is bounded on $\mathrm{L}^2(\mathbb{R})$, we extend H to the whole of $\mathrm{L}^2(\mathbb{R})$ by defining $\widehat{Hf}(\xi) = -\imath \, \mathrm{sgn}(\xi)\hat{f}(\xi)$ almost everywhere in $\mathbb{R}$. Then, this extension agrees on $\mathscr{S}(\mathbb{R})$ and by Plancherel Theorem, $\|Hf\|_2 = \|f\|_2$. $\qquad\square$

Proposition 7.2.3. $H \in \mathrm{CZO}_1$.

Proof. Let $K \in \mathrm{CZK}_1$ be defined by

$$K(x,y) = \frac{1}{\pi(x-y)},$$

when $x \neq y$. Fix $f \in \mathrm{L}^2(\mathbb{R})$ with spt f compact. Then, fix $x \in {}^{\mathrm{c}}(\mathrm{spt}\, f)$ and choose r such that $B(x,r) \cap \mathrm{spt}\, f = \varnothing$. So, there exists a sequence $\varphi_n \in \mathrm{C}_{\mathrm{c}}^\infty(\mathbb{R})$ such that spt $\varphi_n \cap B(x,r) = \varnothing$ and $\varphi_n \to f$ in $\mathrm{L}^2(\mathbb{R}^n)$. Then, for every $z \in B(x,r)$,

$$H\varphi_n(z) = \int_{\mathbb{R}} K(z,y)\varphi_n(y) \, d\mathscr{L}(y) \to \int_{\mathbb{R}} K(z,y)f(y) \, d\mathscr{L}(y)$$

and $H\varphi_n \to Hf$ in $\mathrm{L}^2(\mathbb{R}^n)$. Covering ${}^{\mathrm{c}}(\mathrm{spt}\, f)$ with countably many such balls, we conclude that

$$Hf(x) = \int_{\mathbb{R}^n} K(x,y)f(y) \, d\mathscr{L}(y)$$

almost everywhere $x \in {}^{\mathrm{c}}(\mathrm{spt}\, f)$. Therefore $H \in \mathrm{CZO}_1$. $\qquad\square$

The Hilbert Transform comes from Complex Analysis. Let $f \in \mathrm{C}_{\mathrm{c}}^\infty(\mathbb{R})$ and take the Cauchy extension f to $\mathbb{C} \setminus \mathbb{R}$. That is,

$$F(z) = \frac{1}{2\pi\imath} \int_{\mathbb{R}} \frac{f(t)}{z-t} \, d\mathscr{L}(t)$$

where $z = x + \imath y$, $y \neq 0$. It is an easy fact that F is holomorphic on $\mathbb{C} \setminus \mathbb{R}$. But $\mathbb{C} \setminus \mathbb{R}$ is not connected. So,

$$\lim_{y \to 0^+} F(x \pm \imath y) = \frac{1}{2}(\mp f(x) + \frac{1}{\imath} Hf(x))$$

and consequently

$$\frac{1}{\imath} Hf(x) = \lim_{y \to 0^+} F(x + \imath y) + F(x - \imath y)$$

and

$$f(x) = \lim_{y \to 0^+} F(x - \imath y) - F(x + \imath y).$$

We have the following Theorem of M.Riesz:

Theorem 7.2.4 (Boundedness of the Hilbert Transform). *H has a bounded extension to $\mathrm{L}^p(\mathbb{R})$ for $1 < p < \infty$.*

Corollary 7.2.5. *Let $F_\pm(x) = \lim_{y \to 0+} F(x \pm \imath y)$. Then the decomposition $f = F_- - F_+$ is topological in $\mathrm{L}^p(\mathbb{R}^n)$. That is $\|f\|_p \simeq \|F_+\|_p + \|F_-\|_p$.*

Remark 7.2.6. *When f is real valued, $-\frac{1}{2}Hf$ is the imaginary part of F_+.*

7.2.2 Riesz Transforms

Motivated by the symbol side of the Hilbert Transform, we define operators R_j for $j = 1, \ldots, n$ on $\mathbb{R}^n$.

Definition 7.2.7 (Riesz Transform). *Define $R_j : \mathrm{L}^2(\mathbb{R}^n) \to \mathrm{L}^2(\mathbb{R}^n)$ by*

$$(R_j f)\check{}(\xi) = -\imath \frac{\xi_j}{|\xi|} \hat{f}(\xi)$$

for $j = 1, \ldots, n$.

We note that by Plancherel's Theorem, R_j is well defined and in particular $\|R_j f\|_2 \le \|f\|_2$.

Proposition 7.2.8. $R_j \in \mathrm{CZO}_1$.

Proof. Consider

$$K_j(x) = \mathrm{p.\,v.}\, c_n \frac{x_j}{|x|^{n+1}}$$

for some $c_n > 0$. Then $K_j \in \mathscr{S}'(\mathbb{R}^n)$. If we can show that for appropriate c_n,

$$\widehat{K_j}(\xi) = -\imath \frac{\xi_j}{|\xi|}$$

in $\mathscr{S}'(\mathbb{R}^n)$, by the same argument as for the Hilbert Transform,

$$R_j f(x) = c_n \int_{\mathbb{R}^n} \frac{x_j - y_j}{|x - y|^{n+1}} f(y) \, d\mathscr{L}(y)$$

for all $f \in \mathrm{L}^2(\mathbb{R}^n)$ with spt f compact and for almost every $x \in {}^c(\mathrm{spt}\, f)$.

We compute the Fourier Transform of K_j. Fix $\varphi \in \mathscr{S}(\mathbb{R}^n)$. Then,

$$\langle \widehat{K_j}, \varphi \rangle = \langle K_j, \hat{\varphi} \rangle = \lim_{\varepsilon \to 0} c_n \int_{\{x : \varepsilon < |x| < \varepsilon^{-1}\}} \frac{x_j}{|x|^{n+1}} \int_{\mathbb{R}^n} e^{-\imath x \cdot \xi} \varphi(\xi) \, d\mathscr{L}(\xi) \, d\mathscr{L}(x).$$

For $\xi \neq 0$, let

$$I_\varepsilon = c_n \int_{\{x : \varepsilon < |x| < \varepsilon^{-1}\}} \frac{x_j}{|x|^{n+1}} e^{-\imath x \cdot \xi} \, d\mathscr{L}(x).$$

As before, we show that $|I_\varepsilon|$ is uniformly bounded in ξ and ε and that

$$I_\varepsilon \to -\imath \frac{\xi_j}{|\xi|}$$

64

as $\varepsilon \to 0$.

As previously, write $e^{-ix\cdot\xi} = \cos(x\cdot\xi) - i\sin(x\cdot\zeta)$. We only need to regard the imaginary part. By a change of variables, let $w = \frac{\xi}{|\xi|}$ and $x = |\xi| y$. Then,

$$I_\varepsilon = -i c_n \int_{\left\{x: \frac{\varepsilon}{|\xi|} < |x| < \frac{1}{|\xi|\varepsilon}\right\}} \frac{y_j}{|y|^{n+1}} \sin(y\cdot w) \, d\mathscr{L}(y)$$

since the Jacobian factor of the change of variables is cancelled by the homogeneity of $\frac{x_j}{|x|^{n+1}}$.

We change variables again, this time to polar coordinates. Let $y = r\theta$, for $r > 0$ and $\theta \in S^{n-1}$. Then,

$$I_\varepsilon = -i c_n \int_{S^{n-1}} \theta_j \left(\int_{\frac{\varepsilon}{|\xi|}}^{\frac{1}{\varepsilon|\xi|}} \frac{\sin(r\theta\cdot w)}{r} \, dr \right) d\sigma(\theta)$$

where $d\sigma$ is the surface measure on S^{n-1}. So, $|I_\varepsilon|$ is uniformly bounded since

$$\left| \int_{\frac{\varepsilon}{|\xi|}}^{\frac{1}{\varepsilon|\xi|}} \frac{\sin(r\theta\cdot w)}{r} \, dr \right|$$

is uniformly bounded in $\varepsilon, |\xi|$ and θ. Furthermore,

$$\int_{\frac{\varepsilon}{|\xi|}}^{\frac{1}{\varepsilon|\xi|}} \frac{\sin(r\theta\cdot w)}{r} \, dr \to \frac{\pi}{2} \operatorname{sgn}(\theta\cdot w)$$

as $\varepsilon \to 0$ and so

$$I_\varepsilon \to -i c_n \frac{\pi}{2} \int_{S^{n-1}} \theta_j \operatorname{sgn}(w\cdot w) \, d\sigma(\theta).$$

Write

$$a_j = \int_{S^{n-1}} \theta_j \operatorname{sgn}(\theta\cdot w) \, d\sigma(\theta)$$

and let

$$a = (a_1, \ldots, a_n) = -i c_n \frac{\pi}{2} \int_{S^{n-1}} ((\theta - (\theta\cdot w)w) + (\theta\cdot w)w) \operatorname{sgn}(\theta\cdot w) \, d\sigma(\theta)$$

$$= -i c_n \frac{\pi}{2} \left(\int_{S^{n-1}} |\theta\cdot w| \, d\sigma(\theta) \right) w.$$

because $(\theta - (\theta\cdot w)w) \operatorname{sgn}(\theta\cdot w)$ is odd in the symmetry with respect to the hyperplane $\{w\}^{\perp}$ and S^{n-1} is invariant under this symmetry. By rotational invariance,

$$\int_{S^{n-1}} |\theta\cdot w| \, d\sigma(\theta) = \int_{S^{n-1}} |\theta_1| \, d\sigma(\theta)$$

and so we define c_n by

$$c_n \frac{\pi}{2} \int_{S^{n-1}} |\theta_1| \, d\sigma(\theta) = 1.$$

Then, it follows that

$$a_j = -i w_j = -i \frac{\xi_j}{|\xi|}$$

and the proof is complete. $\qquad\square$

Theorem 7.2.9. R_j *is bounded on* $\mathrm{L}^p(\mathbb{R}^n)$ *whenever* $1 < p < \infty$.

Corollary 7.2.10 (Application to PDEs). *Let* $\varphi \in \mathscr{S}(\mathbb{R}^n)$. *Then,* $\partial_i \partial_j \varphi = -R_i R_j \Delta \varphi$ *and*

$$\|\partial_i \partial_j \varphi\|_p \leq C(n, p)\|\Delta \varphi\|_p$$

whenever $1 < p < \infty$.

Proof. We note that for all $\xi \in \mathbb{R}^n$,

$$
\begin{aligned}
(\partial_i \partial_j \varphi)\widehat{\ }(\xi) &= (-\imath \xi_i)(-\imath \xi_j)\hat{\varphi}(\xi) \\
&= \left(-\imath \frac{\xi_i}{|\xi|}\right)\left(-\imath \frac{\xi_j}{|\xi|}\right)|\xi|^2 \, \hat{\varphi}(\xi) \\
&= \left(-\imath \frac{\xi_i}{|\xi|}\right)\left(-\imath \frac{\xi_j}{|\xi|}\right)\left(\sum_{j=1}^n \xi_j^2 \hat{\varphi}(\xi)\right) \\
&= \left(-\imath \frac{\xi_i}{|\xi|}\right)\left(-\imath \frac{\xi_j}{|\xi|}\right)(-\Delta \varphi)\widehat{\ }(\xi)
\end{aligned}
$$

and by application of the theorem, the proof is complete. $\qquad\square$

7.2.3 Cauchy Operator

The Cauchy Operator is an example of an operator that is *not* of convolution type.

Identify $\mathbb{R}^2 \simeq \mathbb{C}$. Let $\varphi : \mathbb{R} \to \mathbb{R}$ be a Lipschitz map. That is, there exists an $M > 0$ such that $|\varphi(t) - \varphi(s)| \leq M\,|t - s|$. By Rademacher's Theorem [Fed96, Theorem 3.1.6], φ is differentiable almost everywhere and $\varphi' \in \mathrm{L}^\infty(\mathbb{R})$ with $\|\varphi'\|_\infty \leq M$. Now, let $\Gamma = \{t + \imath\varphi(t) : t \in \mathbb{R}\} \subset \mathbb{C}$.

If f is smooth in a neighbourhood of Γ and has compact support, then whenever $z \notin \Gamma$, define

$$F(z) = \frac{1}{2\pi\imath}\int \frac{f(w)}{z - w}\, dw = \frac{1}{2\pi\imath}\int_{\mathbb{R}} \frac{f(s + \imath\varphi(s))}{z - (s + \imath\varphi(s))}(1 + \imath\varphi'(s))\, ds$$

where $z = Z(t) + \imath\alpha$, $\alpha \in \mathbb{R}^*$ and $Z(t) = t + \imath\varphi(t)$. Fix t. Then,

$$\lim_{\alpha \to 0^\pm} F(Z(t) + \imath\alpha) = \mp\frac{1}{2}f(z(t)) + \mathcal{C}f(z(t))$$

(which are the *Plemelj* formula's - details in [Mey92, Volume 2]) where the Cauchy operator is given by

$$\mathcal{C}f(z(t)) = \lim_{\varepsilon \to 0}\frac{1}{2\pi\imath}\int_{\{s:|z(t)-z(s)|>\varepsilon\}} \frac{f(z(s))}{z(t) - z(s)}z'(s)\, ds = \lim_{\varepsilon \to 0}\frac{1}{2\pi\imath}\int_{\{w\in\Gamma:|z-w|>\varepsilon\}} \frac{f(w)}{z - w}\, dw.$$

Let $\tilde{f}(s) = f(z(s))z'(s)$. Then,

$$\mathcal{C}f(z(t)) = \mathrm{p.\,v.}\int_{\{s:|z(t)-z(s)|>\varepsilon\}} \frac{1}{z(t) - z(s)}\tilde{f}(s)\, ds = T\tilde{f}(t).$$

Theorem 7.2.11 (Coifman, McIntosh, Meyer (1982)). $T \in \mathrm{CZO}_1$ *with kernel*

$$\mathrm{p.\,v.} \frac{1}{z(t) - z(s)} \in \mathrm{CZK}_1.$$

The hard part of the theorem is to show $\|T\tilde{f}\|_2 \leq C\|\tilde{f}\|_2$. As a consequence,

Corollary 7.2.12. *(i)* $\mathcal{C}$ *is bounded on* $\mathrm{L}^2(\Gamma, |dw|)$ *where* $|dw|$ *is the arclength measure,*

(ii) The decomposition

$$f(z) = -\lim_{\alpha \to 0^+} F(z(t) + \imath\alpha) + \lim_{\alpha \to 0^-} F(z(t) + \imath\alpha)$$

is topological in $\mathrm{L}^p(\Gamma, |dw|)$.

These results have important applications in boundary value problems, geometric measure theory and partial differential equations.

Remark 7.2.13. *We emphasise that this operator* $\mathcal{C}$ *is not of convolution type. Unlike in the previous two examples, we cannot employ the Fourier transform or simple techniques.*

7.3 L^p boundedness of CZO_α operators

The L^2 boundedness of CZO_α operators comes for free by definition. It is an interesting question to ask when $T \in \mathrm{CZO}_\alpha$ is a bounded map from $\mathrm{L}^q(\mathbb{R}^n)$ to $\mathrm{L}^p(\mathbb{R}^n)$. But first, we have the following proposition which shows that at least for the Hilbert transform, $p = q$.

Proposition 7.3.1. *Suppose the Hilbert transform* $H : \mathrm{L}^q(\mathbb{R}) \to \mathrm{L}^p(\mathbb{R})$ *for some* $p, q > 1$ *is bounded. Then,* $p = q$.

Proof. Let $f \in \mathrm{L}^q(\mathbb{R}^n)$ and consider the function $g(x) = f(\lambda x)$ for $\lambda > 0$. Then,

$$\lambda^\alpha \|Hf\|_q \leq C\lambda^\beta \|f\|_p$$

and so $\alpha = -\frac{1}{q}, \beta = -\frac{1}{p}$ and $\alpha = \beta$ which implies $p = q$. $\qquad \square$

As a heuristic, we cannot hope to prove L^q to L^p boundedness unless $p = q$.

Definition 7.3.2 (Hörmander kernel). *Let* $K \in \mathrm{L}^1_{\mathrm{loc}}(^c\Delta)$ *and suppose there exists* $C_H > 0$ *such that*

$$\mathrm{esssup}_{(y,y') \in \mathbb{R}^{2n}} \int_{\{x : |x-y| \geq 2|y-y'|\}} \left| K(x, y) - K(x, y') \right| \, d\mathscr{L}(x) \leq C_H.$$

Then, K *is called a Hörmander kernel.*

Remark 7.3.3. *The number* 2 *appearing in the set of integration is irrelevant. This can be replaced by any* $A > 1$ *at the cost of changing* C_H.

Lemma 7.3.4. *(i) Every* CZK_α *kernel is Hörmander .*

(ii) The adjoint of a CZK_α *kernel is Hörmander .*

Proof. The proof of (ii) follows easily from (i) observing that $K \in \mathrm{CZK}_\alpha$ implies $K^* \in \mathrm{CZK}_\alpha$.

We prove (i). Let $K \in \mathrm{CZK}_\alpha$ and so we have that

$$\left| K(x,y) - K(x,y') \right| \le C_\alpha \left(\frac{|y-y'|}{|x-y|} \right)^\alpha \frac{1}{|x-y|^n}$$

whenever $|y-y'| \le \frac{1}{2}|x-y|$ and $x \ne y$. So,

$$\int_{\{x:|x-y|\ge 2|y-y'|\}} \left(\frac{|y-y'|}{|x-y|} \right)^\alpha \frac{1}{|x-y|^n}\, d\mathscr{L}(x)$$

$$= \sum_{j=0}^{\infty} \int_{\{x:2^j 2|y-y'|\le|x-y|\le 2^{j+1}2|y-y'|\}} \left(\frac{|y-y'|}{|x-y|} \right)^\alpha \frac{1}{|x-y|^n}\, d\mathscr{L}(x)$$

$$\le \frac{|y-y'|^\alpha}{2\,|y-y'|^{\alpha+n}} \sum_{j=0}^{\infty} \mathscr{L}\left(B(y, 2^{j+1}\,|y-y'|)\right)$$

$$\le A(\alpha, n).$$

$\square$

We now present the following important and main lemma.

Lemma 7.3.5 (Calderón-Zygmund decomposition). *Let $f \in \mathrm{L}^1(\mathbb{R}^n)$ and $\lambda > 0$. Then there exists a $C(n) > 0$ and a decomposition of $f = g + b$ almost everywhere on $\mathbb{R}^n$ where $g \in \mathrm{L}^\infty(\mathbb{R}^n)$ with $\|g\|_\infty \le C(n)\lambda$, and $b = \sum_{i=1}^{\infty} b_i$ where*

(i) $b_i = 0$ on $^c B_i$ with B_i a ball,

(ii) $\int_{B_i} |b_i|\, d\mathscr{L} \le C(n)\lambda \mathscr{L}(B_i)$,

(iii) $\int_{\mathbb{R}^n} b_i = 0$,

(iv) $\{B_i\}$ have the bounded overlap property

$$\sum_{i=1}^{\infty} \chi_{B_i} \le C(n),$$

(v) $\sum_{i=1}^{\infty} \mathscr{L}(B_i) \le C(n)\frac{1}{\lambda}\|f\|_1$.

Remark 7.3.6. *(i) The constant $C(n)$ depends only on the dimension n.*

(ii) *Note that $\sum_{i=1}^{\infty} \|b_i\|_1 \le C(n)\lambda \sum_{i=1}^{\infty} \mathscr{L}(B_i) \le C(n)^2\|f\|_1$ which shows that $\sum_{i=1}^{\infty} b_i$ converges in L^1. Hence, $b \in \mathrm{L}^1(\mathbb{R}^n)$ with $\|b\|_1 \le C(n)^2\|f\|_1$.*

(iii) The fact that $g \in L^\infty(\mathbb{R}^n) \cap L^1(\mathbb{R}^n)$ implies $g \in L^p(\mathbb{R}^n)$ for all $p \in [1, \infty]$. In the case of $p = 2$,

$$\|g\|_2 \leq \sqrt{\|g\|_1 \|g\|_\infty} \leq \sqrt{(1 + C(n))^2 C(n)} \sqrt{\lambda \|f\|_1}.$$

Proof of the Calderón-Zygmund decomposition. Recall that $\mathcal{M}'f$ is the uncentred maximal function of f on balls of $\mathbb{R}^n$. We know that the set $\Omega_\lambda = \{x \in \mathbb{R}^n : \mathcal{M}'f(x) > \lambda\}$ is open and of finite measure by the Maximal Theorem:

$$\mathcal{L}(\Omega_\lambda) \leq \frac{C}{\lambda} \|f\|_1.$$

Also, $\Omega_\lambda \neq \mathbb{R}^n$. Let $\mathcal{E}$ be a Whitney covering of Ω_λ. Set $\left\{ B_i = c_1 \tilde{B}_i : \tilde{B}_i \in \mathcal{E} \right\}$ where c_1 is the constant in the Whitney Covering Lemma 2.3.4. Then, (iv) is proved and

$$\sum_{i=1}^\infty \mathcal{L}(B_i) = \int \sum_{i=1}^\infty \chi_{B_i} \, d\mathcal{L} \leq \int C(n) \chi_{\Omega_\lambda} \, d\mathcal{L} \leq C \frac{C(n)}{\lambda} \|f\|_1$$

which proves (v).

We can now take $c \in (0, 1)$ (say, $c = c_1^{-1}$) and so $\{cB_i\}$ are mutually disjoint. Then, construct a partition of unity φ_i so that $\sum_i \varphi_i = 1$ on Ω_λ. Explicitly,

$$\varphi_i = \frac{\chi_{B_i}}{\sum_j \chi_{B_j}}.$$

Now, set

$$b_i = \begin{cases} f\varphi_i - \fint_{B_i} f\varphi_i \, d\mathcal{L} & \text{on } B_i \\ 0 & \text{otherwise} \end{cases}.$$

Since we allow B_i to be closed we (i) is proved and (iii) is apparent from the construction of b_i.

Now, to prove (ii), we note that $\int_{B_i} |b_i| \, d\mathcal{L} \leq 2 \int_{B_i} |f| \, d\mathcal{L}$ and

$$4B_i \cap {}^c\Omega_\lambda = 4c_1 \tilde{B}_i \cap {}^c\Omega \neq \varnothing.$$

Then, $\int_{4B_i} |f| \, d\mathcal{L} \leq \mathcal{M}'f(z) \mathcal{L}(4B_i)$ for all $z \in 4B_i$. Choosing $z \in {}^c\Omega_\lambda$ we observe that $\mathcal{M}'f(z) \leq \lambda$ and so

$$\int_{B_i} |b_i| \, d\mathcal{L} \leq 2\lambda \mathcal{L}(4B_i) \leq 2 \, 4^n \lambda \mathcal{L}(B_i)$$

which establishes (ii).

Define:

$$g = \begin{cases} f & \text{on } {}^c\Omega_\lambda \\ \sum_i (\fint_{B_i} f\varphi_i \, d\mathcal{L}) \chi_{B_i} & \text{on } \Omega_\lambda \end{cases}.$$

Then, on ${}^c\Omega_\lambda$, $f \leq \mathcal{M}'f \leq \lambda$ almost everywhere. On Ω_λ, by invoking the bounded overlap property,

$$\left| \sum_i \left(\fint_{B_i} f\varphi_i \, d\mathcal{L} \right) \chi_{B_i} \right| \leq C(n) \sup_i \left| \fint_{B_i} f\varphi_i \, d\mathcal{L} \right| \leq C(n) \sup_i \fint_{B_i} |f| \, d\mathcal{L} \leq C(n) 4^n \lambda.$$

This completes the proof. $\qquad\square$

Theorem 7.3.7. *Every* $T \in \mathrm{CZO}_\alpha$ *is of weak type* $(1,1)$.

We have the following immediate consequence.

Corollary 7.3.8. *Let* $T \in \mathrm{CZO}_\alpha$. *Then, for all* $p \in (1, \infty)$, T *is strong type* (p, p).

Proof. Since T is weak type $(1,1)$ by the Theorem and strong type $(2,2)$ by definition, we have that T is strong type (p, p) for $p \in (1, 2)$.

Now, note that $T \in \mathrm{CZO}_\alpha$ implies that $T^* \in \mathrm{CZO}_\alpha$ and so T^* has a bounded extension to $\mathrm{L}^p(\mathbb{R}^n)' = \mathrm{L}^{p'}(\mathbb{R}^n)$ for $1 < p' < 2$. Therefore, T has a bounded extension to $\mathrm{L}^p(\mathbb{R}^n)$ for $2 < p < \infty$. $\qquad\square$

Proof of Theorem 7.3.7. Let $f \in \mathrm{L}^1(\mathbb{R}^n) \cap \mathrm{L}^2(\mathbb{R}^n)$ and fix $\lambda > 0$. We show that

$$\mathscr{L}(\{x \in \mathbb{R}^n : |Tf(x)| > \lambda\}) \leq \frac{C}{\lambda}\|f\|_1$$

with C independent of f and λ. Since we only know $Tf(x)$ when $x \notin \mathrm{spt}\, f$, we use the Calderón-Zygmund decomposition to localise. Let $f = g + b$ this decomposition at level λ with the properties of g and b from Lemma 7.3.5. Since $f, g \in \mathrm{L}^2(\mathbb{R}^n)$, we also have $b \in \mathrm{L}^2(\mathbb{R}^n)$. Since $b = \sum_{i=1}^\infty b_i$, with $b_i = (f\varphi_i - \mathrm{m}_{B_i} f\varphi_i)\chi_{B_i}$ we have that this series converges in $\mathrm{L}^2(\mathbb{R}^n)$. So, $Tf = Tg + Tb$ and we estimate by Markov's inequality

$$A = \mathscr{L}\left(\left\{x \in \mathbb{R}^n : |Tg(x)| > \frac{\lambda}{2}\right\}\right) \leq \frac{4}{\lambda^2}\int_{\mathbb{R}^n} |Tg|^2 \, d\mathscr{L} \leq \frac{4}{\lambda^2}\|T\|_{\mathcal{L}(\mathrm{L}^2(\mathbb{R}^n))}\|g\|_2^2$$

and use $\|g\|_2^2 \leq C(n)\lambda\|f\|_1$.

Now, $T(b) = T(\sum_{i=1}^\infty b_i) = \sum_{i=1}^\infty T(b_i)$ with the series on the right converging in $\mathrm{L}^2(\mathbb{R}^n)$ and $|T(b)| \leq \sum_i |T(b_i)|$ almost everywhere. So, with $c > 1$ to be chosen later,

$$
\begin{aligned}
B = \mathscr{L}\left(\left\{x \in \mathbb{R}^n : |Tb(x)| > \frac{\lambda}{2}\right\}\right) &\leq \mathscr{L}\left(\left\{x \in \mathbb{R}^n : \sum_{i=1}^\infty |Tb_i(x)| > \frac{\lambda}{2}\right\}\right) \\
&\leq \mathscr{L}(\cup_j cB_j) + \mathscr{L}\left(\left\{x \in \mathbb{R}^n \setminus (\cup_j cB_j) : \sum_{j=1}^\infty |Tb_i(x)| > \frac{\lambda}{2}\right\}\right) \\
&\leq \sum_{j=1}^\infty \mathscr{L}(cB_j) + \frac{1}{\lambda}\int_{\mathbb{R}^n \setminus (\cup_j cB_j)} \sum_{i=1}^\infty |Tb_i| \, d\mathscr{L} \\
&\leq c^n \frac{C(n)}{\lambda}\|f\|_1 + \frac{1}{\lambda}\int_{\mathbb{R}^n \setminus (\cup_j cB_j)} \sum_{i=1}^\infty |Tb_i| \, d\mathscr{L}.
\end{aligned}
$$

Consequently, it is enough to prove that

$$\sup_i \int_{\mathbb{R}^n \setminus cB_i} |Tb_i| \, d\mathscr{L} \leq C(T)\|b_i\|_1$$

70

since $\|b_i\|_1 \leq C(n)\lambda \mathscr{L}(B_i)$ which gives

$$\sum_{i=1}^{\infty} \frac{1}{\lambda} \|b_i\|_1 \leq C(n) \sum_{i=1}^{\infty} \mathscr{L}(B_i) \leq \frac{C(n)^2}{\lambda} \|f\|_1.$$

We note that for almost everywhere $x \in \mathbb{R}^n \setminus cB_i$, $Tb_i(x) = \int_{B_i} K(x,y)b_i(y)\, d\mathscr{L}(y)$. Let y_i be the centre of the ball B_i. Since $\int_{\mathbb{R}^n} b_i \, d\mathscr{L} = 0$, for almost all $x \in \mathbb{R}^n \setminus cB_i$,

$$Tb_i(x) = \int_{B_i} (K(x,y) - K(x,y_i))b_i(y)\, d\mathscr{L}(y).$$

We choose $c = 2$ since $2\,|y - y_i| \leq 2\,\mathrm{rad}\,B_i \leq |x - y|$ and

$$\int_{\mathbb{R}^n \setminus 2B_i} |Tb_i|\, d\mathscr{L} \leq \int_{y \in B_i} |b_i(y)| \left(\int_{\{x:|x-y|\geq 2|y-y_i|\}} |K(x,y) - K(x,y_i)|\, d\mathscr{L}(x) \right) d\mathscr{L}(y)$$

$$\leq \int_{\mathbb{R}^n} |b_i(y)|\, C_H(K)\, d\mathscr{L}(y) \leq C_H(K)\|b_i\|_1$$

where $C_H(K)$ is the Hörmander constant associated with K. Taking an infimum on the right hand side, we have

$$\int_{\mathbb{R}^n \setminus 2B_i} |Tb_i|\, d\mathscr{L} \leq \|K\|_{\mathrm{CZK}_\alpha} \|b_i\|_1.$$

The sum $A + B$ gives us the desired conclusion with constant $C \leq C(n)(\|T\|_{\mathcal{L}(\mathrm{L}^2(\mathbb{R}^n))} + \|K\|_{\mathrm{CZK}_\alpha} = C(n)\|T\|_{\mathrm{CZO}_\alpha}$.

For a general $f \in \mathrm{L}^1(\mathbb{R}^n)$ let $f_k \to f$ be a sequence which converges in $\mathrm{L}^1(\mathbb{R}^n)$ with each $f_k \in \mathrm{L}^2(\mathbb{R}^n)$. Without loss of generality, assume that $f_k \to f$ almost everywhere (since we can pass to a subsequence). The weak type $(1,1)$ condition gives that Tf_k is Cauchy in measure and call $\tilde{T}f$ the limit. This exists almost everywhere and $\tilde{T}f \in \mathrm{L}^{1,\infty}(\mathbb{R}^n)$. Furthermore,

$$\tilde{T}(f)(x) = \int_{\mathbb{R}^n} K(x,y)f(y)\, d\mathscr{L}(y)$$

for almost every $x \in {}^c(\mathrm{spt}\, f)$ with $\mathrm{spt}\, f$ compact. $\qquad \square$

Remark 7.3.9. *It would also suffice to prove for general f in the previous Theorem by noting $\mathrm{L}^1(\mathbb{R}^n) \cap \mathrm{L}^2(\mathbb{R}^n)$ is dense in $\mathrm{L}^1(\mathbb{R}^n)$ and that $T : \mathrm{L}^1(\mathbb{R}^n) \cap \mathrm{L}^2(\mathbb{R}^n) \to \mathrm{L}^{1,\infty}(\mathbb{R}^n)$ is bounded. Since $\mathrm{L}^{1,\infty}(\mathbb{R}^n)$ is complete, T extends to a bounded map $\tilde{T} : \mathrm{L}^1(\mathbb{R}^n) \to \mathrm{L}^{1,\infty}(\mathbb{R}^n)$.*

Example 7.3.10. *We note that*

$$H(\chi_{[0,1]})(x) = -\frac{1}{\pi} \ln \left| \frac{x-1}{x} \right|$$

whenever $x \notin [0,1]$.

This example is of importance because H is a CZO, $\chi_{[0,1]} \in \mathrm{L}^1(\mathbb{R})$ but $H(\chi_{[0,1]}) \notin \mathrm{L}^1(\mathbb{R})$.

7.4 CZO and H^1

A natural question to ask is: what subspace of L^1 should we choose so that a CZO$_\alpha$ maps that space back into L^1?

Theorem 7.4.1. *Let $T \in$ CZO$_\alpha$. Then, T induces a bounded operator H$^1 \to$ L$^1(\mathbb{R}^n)$.*

Corollary 7.4.2. *Let $T \in$ CZO$_\alpha$. Then, T extends to a bounded operator from L$^\infty(\mathbb{R}^n)$ to BMO.*

Proof. Let $f \in$ L$^\infty(\mathbb{R}^n)$ and $g \in$ H^1. Then, $Lg = \langle f, T^{\mathrm{tr}} g \rangle$ is a linear functional on H^1 satisfying

$$|\langle f, T^{\mathrm{tr}} g \rangle| = \left| \int_{\mathbb{R}^n} f T^{\mathrm{tr}} g \, d\mathscr{L} \right| \leq \|f\|_\infty \|T^{\mathrm{tr}}\|_{\mathcal{L}(\mathrm{H}^1, \mathrm{L}^1(\mathbb{R}^n))} \|g\|_{\mathrm{H}^1}.$$

By duality, there exists a $\beta \in$ BMO such that $L = L_\beta$. Define $Tf = \beta$, with β identified with L_β. $\qquad\square$

Remark 7.4.3. *(i) This was originally proved directly, without alluding to duality. See [Ste93].*

(ii) We apply Tf to p-atoms. Let $a \in \mathscr{A}^p$. Then,

$$\langle Tf, a \rangle = L_\beta(a) = \int_{\mathbb{R}^n} \beta a \, d\mathscr{L}.$$

Let $B = B(y_B, r_B) = \mathrm{spt}\, a$. Then,

$$
\begin{aligned}
\langle Tf, a \rangle &= \int_{\mathbb{R}^n} f T^{\mathrm{tr}}(a) \, d\mathscr{L} \\
&= \int_{2B} f T^{\mathrm{tr}}(a) \, d\mathscr{L} + \int_{\mathbb{R}^n \setminus 2B} f T^{\mathrm{tr}}(a) \, d\mathscr{L} \\
&= \int_{\mathbb{R}^n} T(f\chi_{2B}) a \, d\mathscr{L} + \int_{\mathbb{R}^n \setminus 2B} f(y) \left(\int_{\mathbb{R}^n} K(x,y) a(x) \, d\mathscr{L}(x) \right) d\mathscr{L}(y) \\
&= \int_{\mathbb{R}^n} T(f\chi_{2B}) a \, d\mathscr{L} + \\
&\qquad \int_{\mathbb{R}^n} \left(\int_{\mathbb{R}^n \setminus 2B} f(y) (K(x,y) - K(y_B, y)) \, d\mathscr{L}(y) \right) a(x) \, d\mathscr{L}(x) \, d\mathscr{L}(y)
\end{aligned}
$$

by the application of Fubini. So, on B there exists a constant C_B such that

$$\beta(y) = T(f\chi_{2B})(y) + \int_{\mathbb{R}^n \setminus 2B} f(y)(K(x,y) - K(y_B, y)) \, d\mathscr{L}(y) + C_B.$$

Proof of Theorem 7.4.1. We show that whenever $a \in \mathscr{A}^\infty$, then $Ta \in$ L$^1(\mathbb{R}^n)$ with $\|Ta\|_1 \leq C(n,T)$. We automatically have $Ta \in$ L$^2(\mathbb{R}^n)$ since $a \in$ L$^\infty(\mathbb{R}^n)$ and spt $a \subset B$ a ball. Then, since $\|a\|_2 \leq \frac{1}{\mathscr{L}(B)^{\frac{1}{2}}}$,

$$\int_{2B} |Ta| \, d\mathscr{L} \leq \mathscr{L}(2B)^{\frac{1}{2}} \|Ta\|_{\mathrm{L}^2(2B)} \leq \mathscr{L}(2B)^{\frac{1}{2}} \|T\|_{\mathcal{L}(\mathrm{L}^2(\mathbb{R}^n))} \|a\|_2 \leq 2^{n/2} \|T\|_{\mathcal{L}(\mathrm{L}^2(\mathbb{R}^n))}.$$

As in the proof of Theorem 7.3.7,

$$\int_{\mathbb{R}^n \setminus 2B} |Ta| \, d\mathscr{L} \leq C(n) \|K\|_{\mathrm{CZO}_\alpha} \|a\|_1.$$

Since $\mathrm{H}^1 \subset \mathrm{L}^1(\mathbb{R}^n)$, $Tf \in \mathrm{L}^{1,\infty}(\mathbb{R}^n)$ for every $f \in \mathrm{H}^1$. So, fix $f \in \mathrm{H}^1$ and pick a representation: $f = \sum_{j=1}^\infty \lambda_j a_j$ where $\sum_{j=1}^\infty |\lambda_j| \leq 2\|f\|_{\mathrm{H}^1}$ with $a_j \in \mathscr{A}^\infty$. This series converges almost everywhere in L^1 to f. Thus, $T\left(\sum_{j=1}^\infty \lambda_j a_j\right) = Tf$ almost everywhere. Also,

$$\sum_{j=1}^\infty \|\lambda_j Ta_j\|_1 \leq \sum_{j=1}^\infty |\lambda_j| C(n,T) \leq 2\|f\|_{\mathrm{H}^1} C(n,T).$$

Thus, $\sum_{j=1}^\infty \lambda_j Ta_j$ converges in L^1 and hence,

$$\sum_{j=1}^\infty \lambda_j Ta_j = T\left(\sum_{j=1}^\infty \lambda_j a_j\right)$$

holds almost everywhere (remark that the equality for finite sums is trivial). Hence $Tf \in \mathrm{L}^1(\mathbb{R}^n)$ and $\|Tf\|_1 \leq 2C(n,T)\|f\|_{\mathrm{H}^1}$. $\qquad\square$

Proposition 7.4.4. *Let $T \in \mathrm{CZO}_\alpha$. Then, $T1$ is defined as a BMO function.*

Proof. Follows easily from the fact that $1 \in \mathrm{L}^\infty(\mathbb{R}^n)$, and $T : \mathrm{L}^\infty(\mathbb{R}^n) \to \mathrm{BMO}$ is bounded. $\qquad\square$

Remark 7.4.5. *To compute $T1$, use the formula for Tf on each ball B for $f = 1$.*

Corollary 7.4.6. *Let $T \in \mathrm{CZO}_\alpha$. Then T maps H^1 to H^1 if and only if $T^{\mathrm{tr}}1 = 0$ (in BMO).*

Before we prove this corollary, we need the following lemmas.

Lemma 7.4.7. *Let $T \in \mathrm{CZO}_\alpha$ with associated kernel $K \in \mathrm{CZK}_\alpha$ and $a \in \mathscr{A}^\infty$ with spt $a \subset B = B(y_B, r_B)$. For each $j \in \mathbb{N}$ with $j \geq 1$, let $C_j(B) = 2^{j+1}B \setminus 2^j B$. Then, for all $x \in C_j(B)$,*

$$|Ta(x)| \leq \|K\|_{\mathrm{CZK}_\alpha} 2^{-j(n+\alpha)} r_B^{-n}.$$

Proof. We compute and use the α regularity of K,

$$|Ta(x)| \leq \|K\|_{\mathrm{CZK}_\alpha} \int_{y \in B} \left(\frac{|y - y_B|}{|x - y_B|}\right)^\alpha \frac{1}{|x - y_B|^n} |a(y)| \, d\mathscr{L}(y)$$

$$\leq \|K\|_{\mathrm{CZK}_\alpha} r_B^\alpha \frac{1}{(2^j r_B)^{n+\alpha}} \int_{y \in B} |a(y)| \, d\mathscr{L}(y)$$

and the result follows since

$$\int_{y \in B} |a(y)| \, d\mathscr{L}(y) \leq 1.$$

$\qquad\square$

Lemma 7.4.8. *Let $m : \mathbb{R}^n \to \mathbb{C}$ and $B = B(y_B, r_B)$ a ball such that*

1. $\int_{2B} |m|^2 \, d\mathscr{L} \leq \frac{C}{\mathscr{L}(B)}$,

2. *For every $j \in \mathbb{N}$, $j \geq 1$, and $x \in C_j(B) = 2^{j+1}B \setminus 2^j B$ we have*

$$|m(x)| \leq C 2^{-j(n+\alpha)} r_B^{-n}.$$

Then, $m \in \mathrm{H}^1$ and $\|m\|_{\mathrm{H}^1}$ does not exceed a constant depending on n, $\|K\|_{\mathrm{CZK}_\alpha}$ and $\alpha > 0$.

The proof is left as an exercise.

Proof of Corollary 7.4.6. Let a be an ∞-atom with spt $a \subset B$. By Lemma 7.4.7 and previously, we have shown that a satisfies the hypothesis of Lemma 7.4.8. We conclude that $Ta \in \mathrm{H}^1$ with a *uniform* norm (with respect to a) if and only if $\int_{\mathbb{R}^n} Ta \, d\mathscr{L} = 0$. But, $\int_{\mathbb{R}^n} Ta \, d\mathscr{L} = 0$ for all atoms $a \in \mathscr{A}^\infty$ if and only if $T^{\mathrm{tr}} 1 = 0$. $\qquad\square$

7.5 Mikhlin multiplier Theorem

Definition 7.5.1 (Fourier multiplier operator). *Let $m \in \mathrm{L}^\infty(\mathbb{R}^n)$. Define the Fourier multiplier operator operator $T_m : \mathrm{L}^2(\mathbb{R}^n) \to \mathrm{L}^2(\mathbb{R}^n)$ associated to m by*

$$(T_m f)\hat{} = m \hat{f}.$$

Remark 7.5.2. *By Plancherel theorem, $T_m \in \mathcal{L}(\mathrm{L}^2(\mathbb{R}^n))$ and $\|T_m\|_{\mathcal{L}(\mathrm{L}^2(\mathbb{R}^n))} = \|m\|_\infty$.*

We want to consider when such an operator is bounded on $\mathrm{L}^p(\mathbb{R}^n)$ for $p \neq 2$. Such operators arise naturally - for instance when studying PDE with constant or smooth coefficients. Also, we have the following important example.

Example 7.5.3 (The Riesz Transform). *Consider the Fourier multiplier*

$$m(\xi) = -\imath \frac{\xi_j}{|\xi|}.$$

Then, $T_m = R_j$, the jth Riesz transform.

Theorem 7.5.4 (Mikhlin Multipler Theorem). *Let $m \in \mathrm{L}^\infty(\mathbb{R}^n)$. Assume that $m \in C^\infty(\mathbb{R}^n \setminus \{0\})$ and for all $\alpha \in \mathbb{N}^n$ there exists a $C_\alpha < \infty$ such that*

$$\left| \frac{\partial^{|\alpha|}}{\partial \xi^\alpha} m(\xi) \right| \leq \frac{C_\alpha}{|\xi|^{|\alpha|}}$$

when $\xi \neq 0$, and where $|\alpha| = \alpha_1 + \cdots + \alpha_n$. Then, $T_m \in \mathrm{CZO}_1$, hence T_m has a bounded extension to $\mathrm{L}^p(\mathbb{R}^n)$ for all $p \in (1, \infty)$.

Remark 7.5.5. *Note that T_m (and T_m^{tr}) is bounded on H^1.*

Proof. It is enough to show that T_m has an associated $K \in \mathrm{CZK}_1$. We use the *Littlewood-Paley decomposition*. The idea is to split up $\mathbb{R}^n$ such that $\frac{1}{|\xi|^\alpha}$ is essentially constant on each part.

Take $w \in C^\infty(\mathbb{R}_+, [0,1])$ with spt $w \subset [\frac{1}{2}, 4]$ and $w = 1$ on $[1,2]$. Define:

$$W(t) = \frac{w(t)}{\sum_{j \in \mathbb{Z}} w(2^{-j}t)}$$

for $t > 0$. Observe that $1 \le \sum_{j \in \mathbb{Z}} w(2^{-j}t) \le 4$ for all $t > 0$ and $t \mapsto \sum_{j \in \mathbb{Z}} w(2^{-j}t) \in C^\infty(0, \infty)$. In fact $W \in C^\infty(0, \infty) \cap L^\infty(\mathbb{R}^n)$,

$$\left| \frac{d^k}{dt^k} W(t) \right| \le C_k$$

for all $t \in (0, \infty)$. Define $\varphi : \mathbb{R}^n \to [0, \infty]$ by $\varphi(\xi) = W(|\xi|)$. Then, $\varphi \in C^\infty(\mathbb{R}^n \setminus \{0\})$, radial bounded and

$$\mathrm{spt}\, \varphi \subset C_0 = \left\{ \xi \in \mathbb{R}^n : \frac{1}{2} \le |\xi| \le 4 \right\}$$

for all $\alpha \in \mathbb{N}^n$, and

$$\left\| \frac{\partial^{|\alpha|}}{\partial \xi^\alpha} \varphi \right\|_\infty \le C(n, \alpha).$$

The $\left\{ \varphi_j(\xi) = \varphi(2^{-j}\xi) \right\}$ form a partition of unity: $\sum_{j \in \mathbb{Z}} \varphi_j(\xi) = 1$, $\xi \in \mathbb{R}^n \setminus \{0\}$.

We want to compute $k = \check{m}$ in $\mathscr{S}'(\mathbb{R}^n)$ and show that $|k(x)| \le \frac{c}{|x|^n}$ and $|\nabla k(x)| \le \frac{C}{|x|^{n+1}}$ ($x \ne 0$). This implies the Schwartz kernel of T_m will be given by $k(x-y)$ (in $\mathscr{S}'(\mathbb{R}^{2n})$) and its restriction to $^c\Delta$ is CZK_1.

Let $\psi \in \mathscr{S}(\mathbb{R}^n)$ then

$$
\begin{aligned}
\langle k, \psi \rangle &= \langle \check{m}, \psi \rangle \\
&= \langle m, \check{\psi} \rangle \\
&= \int_{\mathbb{R}^n} m(\xi) \check{\psi}(\xi) \, d\mathscr{L}(\xi) \\
&= \sum_{j \in \mathbb{Z}} \int_{\mathbb{R}^n} m(\xi) \varphi_j(\xi) \check{\psi}(\xi) \, d\mathscr{L}(\xi) \\
&= \sum_{j \in \mathbb{Z}} \int_{\mathbb{R}^n} m(\xi) \varphi_j(\xi) \left(\frac{1}{(2\pi)^n} \int_{\mathbb{R}^n} e^{ix \cdot \xi} \psi(x) \, d\mathscr{L}(x) \right) d\mathscr{L}(\xi) \\
&= \sum_{j \in \mathbb{Z}} \langle k_j, \psi \rangle
\end{aligned}
$$

where

$$\langle k_j, \psi \rangle = \frac{1}{(2\pi)^n} \int_{\mathbb{R}^n} \left(\int_{\mathbb{R}^n} e^{ix \cdot \xi} m(\xi) \varphi_j(\xi) \, d\mathscr{L}(\xi) \right) \psi(x) \, d\mathscr{L}(x).$$

In fact, $m\varphi_j \in L^\infty(\mathbb{R}^n)$ with compact support, so k_j identifies with a bounded C^∞ function and we write

$$k_j(x) = \frac{1}{(2\pi)^n} \int_{\mathbb{R}^n} m(\xi) \varphi_j(\xi) e^{ix \cdot \xi} \, d\mathscr{L}(\xi)$$

for $x \in \mathbb{R}^n$. Also, $k = \sum_{j \in \mathbb{Z}} k_j$ in $\mathscr{S}'(\mathbb{R}^n)$.

We estimate k_j. First,

$$|k_j(x)| \leq \|m\|_\infty \int_{\mathbb{R}^n} |\varphi_j(\xi)| \ d\mathscr{L}(\xi) = \|m\|_\infty \|\varphi_j\|_1 = \|m\|_\infty \|\varphi_1\| 2^{jn}.$$

Then, for x large,

$$k_j(x) = \frac{(-1)^{|\alpha|}}{(2\pi)^n} \int_{\mathbb{R}^n} \frac{\partial^{|\alpha|}}{\partial \xi^\alpha} (m\varphi_j)(\xi) \frac{e^{\imath x \cdot \xi}}{(\imath x)^\alpha} \ d\mathscr{L}(\xi).$$

We cover $\mathbb{R}^n$ by conical sectors. Then, in the first sector, $|x| \simeq |x_1|$. So, choose α such that $\alpha_j = 0$ for $j > 1$ and

$$|k_j(x)| \leq C 2^{-j\alpha} 2^{jn} \left\| \frac{\partial^{|\alpha|}}{\partial \xi^\alpha} \varphi \right\|_1$$

where C depends on the sector. Now, we cover $\mathbb{R}^n$ by a finite number of sectors to obtain

$$|k_j(x)| \leq C_M \frac{2^{jn}}{(1 + 2^j |x|)^M}$$

for all $M \in \mathbb{N}$. Then, for $x \neq 0$,

$$|k(x)| \leq \sum_{j \in \mathbb{Z}} |k_j(x)| \leq \frac{C}{|x|^n}.$$

A repetition of this argument with ∇k in place of k yields the desired estimate for ∇k. $\square$

7.6 Littlewood-Paley Theory

We extend Calderón-Zygmund Operators to a Hilbert space valued functions. In this section, we let $\mathscr{H}$ be a separable complex Hilbert space and we denote its norm by $|\cdot|_{\mathscr{H}} = \sqrt{\langle \cdot, \cdot \rangle_{\mathscr{H}}}$.

We require the following notions.

Definition 7.6.1 (Strongly measurable). *Let $f : \mathbb{R}^n \to \mathscr{H}$. Then, we say that f is strongly measurable if given an orthonormal Schauder basis $\{e_i\}$ of $\mathscr{H}$, the Fourier coefficients $\langle f, e_i \rangle : \mathbb{R}^n \to \mathbb{C}$ are measurable.*

Definition 7.6.2 ($\mathrm{L}^p(\mathbb{R}^n; \mathscr{H})$). *For $1 \leq p < \infty$, we define the Hilbert valued L^p space denoted by $\mathrm{L}^p(\mathbb{R}^n; \mathscr{H})$ to be the set of strongly measurable functions $f : \mathbb{R}^n \to \mathscr{H}$ such that $\int_{\mathbb{R}^n} |f|^p_{\mathscr{H}} \ d\mathscr{L} < \infty$. Then, the L^p norm is defined as $\|f\|_p = \| \, |f|_{\mathscr{H}} \, \|_p$.*

Similarly, for $p = \infty$, we say that $f \in \mathrm{L}^\infty(\mathbb{R}^n; \mathscr{H})$ if f is strongly measurable and $|f|_{\mathscr{H}} \in \mathrm{L}^\infty(\mathbb{R}^n)$. Then, $\|f\|_\infty = \| \, |f|_{\mathscr{H}} \, \|_\infty$.

A deeper discussion of these ideas can be found in [Yos95].

We also extend the notion of a CZK kernel and CZO operator to a Hilbert valued setting.

Definition 7.6.3 (Hilbert CZK$_\alpha$). *Let $\mathscr{H}_1, \mathscr{H}_2$ be separable Hilbert spaces. Then, we say a K is a Hilbert valued Calderón-Zygmund Kernel if $K(x, y) \in \mathcal{L}(\mathscr{H}_1, \mathscr{H}_2)$ for all $(x, y) \in {}^c\Delta$ and satisfies (i) to (iii) in Definition 7.1.1 with the absolute value replaced by $\mathcal{L}(\mathscr{H}_1, \mathscr{H}_2)$ norm. We denote the set of all such K by $\mathrm{CZK}_\alpha(\mathscr{H}_1, \mathscr{H}_2)$.*

Definition 7.6.4 (Hilbert CZO$_\alpha$). *Let $\mathscr{H}_1, \mathscr{H}_2$ be separable Hilbert spaces. We say that $T \in \mathcal{L}(\mathrm{L}^2(\mathbb{R}^n; \mathscr{H}_1), \mathrm{L}^2(\mathbb{R}^n; \mathscr{H}_2))$ is a Calderón-Zygmund operator of order α if it is associated to a $K \in \mathrm{CZO}_\alpha(\mathscr{H}_1, \mathscr{H}_2)$ by:*

$$Tf(x) = \int_{\mathbb{R}^n} K(x, y) f(y) \, d\mathscr{L}(y)$$

almost everywhere $x \in {}^c(\mathrm{spt}\ f)$ with $f \in \mathrm{L}^2(\mathbb{R}^n; \mathscr{H}_1)$ with $\mathrm{spt}\ f$ compact. We denote the set of all such operators by $\mathrm{CZO}_\alpha(\mathscr{H}_1, \mathscr{H}_2)$.

Theorem 7.6.5. *If $T \in \mathrm{CZO}_\alpha(\mathscr{H}_1, \mathscr{H}_2)$, then T induces a continuous extension to a bounded operator from $\mathrm{L}^p(\mathbb{R}^n; \mathscr{H}_1)$ to $\mathrm{L}^p(\mathbb{R}^n; \mathscr{H}_2)$ for $1 < p < \infty$.*

Proof. Same as the "scalar" case. $\qquad\qquad\qquad\qquad\qquad\qquad\qquad\qquad\qquad\quad\square$

Theorem 7.6.6. *Let $\varphi \in \mathscr{S}(\mathbb{R}^n)$ such that $\mathrm{spt}\ \hat{\varphi} \subset \left\{ \xi \in \mathbb{R}^n : \frac{1}{2} \leq |\xi| \leq 4 \right\}$ and $\hat{\varphi}(\xi) = 1$ if $1 \leq |\xi| \leq 2$. Let $1 < p < \infty$. Then, there exist constants $C_1, C_2 < \infty$ depending only on n, φ and p such that*

$$C_1 \|f\|_p \leq \| \left(\sum_{j \in \mathbb{Z}} |\Delta_j f|^2 \right)^{\frac{1}{2}} \|_p \leq C_2 \|f\|_p \tag{LP}$$

for all $f \in \mathrm{L}^p(\mathbb{R}^n)$ where $\Delta_j f = \varphi_j * f$ with $\varphi_j(x) = 2^{jn} \varphi(2^j x)$.

Remark 7.6.7. *Note that $(\Delta_j f)\hat{} = \widehat{\varphi_j} \hat{f}$, $\widehat{\varphi_j}(\xi) = \hat{\varphi}(\frac{\xi}{2^j})$ and $\mathrm{spt}\ \widehat{\varphi_j} \subset \left\{ \xi : \frac{1}{2} \leq \frac{|\xi|}{2^j} \leq 4 \right\} =: C_j$.*

The pieces Δ_j are "almost-orthogonal" for "good" f, g:

$$\int_{\mathbb{R}^n} \Delta_j f(x) \overline{\Delta_l g}(x) \, d\mathscr{L} = \frac{1}{(2\pi)^n} \int_{\mathbb{R}^n} (\Delta_j f)\hat{}(\xi) \overline{(\Delta_l g)\hat{}}(\xi) \, d\mathscr{L}(\xi) = 0$$

if $|j - l| \geq 3$.

This theorem says how to "pack" the $\Delta_j f$ pieces to recover the L^p norm.

Proof. Define $T : \mathrm{L}^2(\mathbb{R}^n; \mathbb{C}) \to \mathrm{L}^2(\mathbb{R}^n; \ell^2(\mathbb{Z}))$ as the map $f \mapsto (\Delta_j f)_{j \in \mathbb{Z}}$.

(i) $T \in \mathcal{L}(\mathrm{L}^2(\mathbb{R}^n; \mathbb{C}), \mathrm{L}^2(\mathbb{R}^n; \ell^2(\mathbb{Z})))$ because:

$$\int_{\mathbb{R}^n} |Tf(x)|^2_{\ell^2(\mathbb{Z})} \, d\mathscr{L}(x) = \int_{\mathbb{R}^n} \sum_{j \in \mathbb{Z}} |\Delta_j f(x)|^2 \, d\mathscr{L}(x)$$

$$= \sum_{j \in \mathbb{Z}} \frac{1}{(2\pi)^n} \int_{\mathbb{R}^n} |\widehat{\varphi_j}(\xi)|^2 \left| \hat{f}(\xi) \right|^2 \, d\mathscr{L}(\xi)$$

$$= \int_{\mathbb{R}^n} m(\xi) \left| \hat{f}(\xi) \right|^2 \, d\mathscr{L}(\xi)$$

where $m(\xi) = \sum_{j \in \mathbb{Z}} \frac{1}{(2\pi)^n} \left| \hat{\varphi}(\frac{\xi}{2^j}) \right|^2$. Also,

$$m(\xi) \leq \frac{4}{(2\pi)^n} \|\hat{\varphi}\|_\infty^2$$

for all $\xi \neq 0$. Thus,

$$\int_{\mathbb{R}^n} |Tf|_{\ell^2(\mathbb{Z})}^2 \, d\mathscr{L} \leq 4 \|\hat{\varphi}\|_\infty^2 \int_{\mathbb{R}^n} |f|^2 \, d\mathscr{L}.$$

(ii) We have $m(\xi) \geq \frac{1}{(2\pi)^n}$ for all $\xi \neq 0$ since $\hat{\varphi}(\xi) = 1$ for $1 \leq |\xi| \leq 2$. However,

$$\int_{\mathbb{R}^n} |Tf|_{\ell^2(\mathbb{Z})}^2 \, d\mathscr{L} \geq \|f\|_2^2$$

so (LP) holds for $p = 2$.

(iii) We apply Calderón-Zygmund theory to show that $T \in \mathrm{CZO}_\alpha(\mathbb{C}, \ell^2(\mathbb{Z}))$.

Let $K_j(x, y) = \varphi_j(x - y)$ for $j \in \mathbb{Z}$ and $x, y \in \mathbb{R}^n$ and $K(x, y) = (K_j(x, y))_{j \in \mathbb{Z}}$ for $x, y \in {}^c\Delta$. We note that $\|K(x, y)\|_{\mathcal{L}(\mathbb{C}, \ell^2(\mathbb{Z}))}^2 = \sum_{j \in \mathbb{Z}} |K_j(x, y)|^2$. We use the fact that $|\varphi_j(x)| = C_\varphi 2^{jn}(1 + 2^j |x|)^{-M}$ whenever $M > n$.

We split the above sum according to $2^j |x - y| \geq 1$ or < 1 and this gives us

$$\sum_{j \in \mathbb{Z}} |K_j(x, y)|^2 \leq C_{\varphi, n} \frac{1}{|x - y|^{2n}}.$$

Also, $\nabla_x K(x, y) = (\nabla_x K_j(x, y))_{j \in \mathbb{Z}}$ and

$$|\nabla_x K(x, y)| \leq C 2^{j(n+1)} |\nabla \varphi| \left| 2^j (x - y) \right| \leq C_\varphi 2^{j(n+1)} (1 + 2^j |x|)^{-M}$$

where $M > n + 1$. This implies that

$$\|\nabla_x K(x, y)\|_{\mathcal{L}(\mathbb{C}, \ell^2(\mathbb{Z}))} \leq \frac{C_\varphi}{|x - y|^{n+1}}$$

and $\nabla_y K(x, y) = -\nabla_x K(x, y)$.

(iv) Let $f \in \mathrm{C}_c^\infty(\mathbb{R}^n)$ and $g \in \mathrm{C}_c^\infty(\mathbb{R}^n, \ell^2(\mathbb{Z}))$ with $\mathrm{spt}\, f \cap \mathrm{spt}\, g = \varnothing$. First, note that this implies $\mathrm{spt}\, f \cap \mathrm{spt}\, |g| = \varnothing$. Then,

$$\int_{\mathbb{R}^n} Tf(x) \cdot \overline{g(x)} \, d\mathscr{L}(x) = \int_{\mathbb{R}^n} \sum_{j=-\infty}^{\infty} \Delta_j f(x) \cdot \overline{g_j(x)} \, d\mathscr{L}(x)$$

$$= \sum_{j=-\infty}^{\infty} \int_{\mathbb{R}^n} \left(\int_{\mathbb{R}^n} k_j(x, y) f(y) \, d\mathscr{L}(y) \right) \overline{g_j(x)} \, d\mathscr{L}(x)$$

$$= \sum_{j=-\infty}^{\infty} \int_{\mathbb{R}^n} \left(\int_{\mathbb{R}^n} k_j(x, y) g_j(x) \, d\mathscr{L}(y) \right) \overline{f(y)} \, d\mathscr{L}(x)$$

$$= \int_{\mathbb{R}^n} \sum_{j=-\infty}^{\infty} \left(\int_{\mathbb{R}^n} k_j(x, y) \overline{g_j(x)} f(y) \, d\mathscr{L}(y) \right) d\mathscr{L}(x)$$

78

by applying Fubini and by the disjoint supports of f and g. Since $Tf(x) \in L^2(\ell^2)$ and $\int_{\mathbb{R}^n} K(x,y)f(y)$ is an ℓ^2 valued integral, we have that

$$Tf(x) = \int_{\mathbb{R}^n} K(x,y)f(y) \, d\mathscr{L}(y)$$

in $L^2_{\mathrm{loc}}(^{\mathrm{c}}(\mathrm{spt}\, f); \ell^2)$ and hence almost everywhere $x \in {}^{\mathrm{c}}(\mathrm{spt}\, f)$.

(v) We conclude that T extends to an operator from $L^p(\mathbb{R}^n; \mathbb{C})$ to $L^p(\mathbb{R}^n; \ell^2(\mathbb{Z}))$ for $1 < p < \infty$. Hence, $(\sum_{j \in \mathbb{Z}} |\Delta_j f|^2)^{\frac{1}{2}} = |Tf|_{\ell^2(\mathbb{Z})} \in L^p(\mathbb{R}^n; \mathbb{C})$ with $\big\| |Tf|_{\ell^2(\mathbb{Z})} \big\|_p \leq C(p, n, \mathrm{CZO}_1)\|f\|_p$.

(vi) We prove the left hand side of the (LP) estimate. We show that there exists a $\tilde{\varphi} \in \mathscr{S}(\mathbb{R}^n)$ such that for all $\xi \in \mathbb{R}^n \setminus \{0\}$,

$$\sum_{j \in \mathbb{Z}} \hat{\varphi}(\frac{\xi}{2^j}) \hat{\tilde{\varphi}}(\frac{\xi}{2^j}) = 1.$$

Let

$$w(\xi) = \begin{cases} \dfrac{\hat{\varphi}(\xi)}{\sum_{j \in \mathbb{Z}} \left| \hat{\varphi}(\frac{\xi}{2^j}) \right|^2} & \xi \in \mathbb{R}^n \setminus \{0\} \\ 0 & \xi = 0 \end{cases}$$

and note that $w \in C^\infty(\mathbb{R}^n)$, with $\mathrm{spt}\, w \subset \mathrm{spt}\, \hat{\varphi}$. Define $\tilde{\varphi} = \check{w} \in \mathscr{S}(\mathbb{R}^n)$. By construction, $\sum_{j \in \mathbb{Z}} \hat{\varphi}(\frac{\xi}{2^j}) \hat{\tilde{\varphi}}(\frac{\xi}{2^j}) = 1$.

Let $f \in L^p(\mathbb{R}^n; \mathbb{C}) \cap L^2(\mathbb{R}^n; \mathbb{C})$ and $g \in C_c^\infty(\mathbb{R}^n)$. Then,

$$\int_{\mathbb{R}^n} f(x)\overline{g(x)} \, d\mathscr{L}(x) = \frac{1}{(2\pi)^n} \int_{\mathbb{R}^n} \hat{f}(\xi)\overline{\hat{g}(\xi)} \, d\mathscr{L}(\xi)$$

$$= \sum_{j \in \mathbb{Z}} \int_{\mathbb{R}^n} \hat{\varphi}_j(\xi)\hat{f}(\xi) \cdot \overline{\hat{\tilde{\varphi}}(\xi)g(\xi)} \, d\mathscr{L}(\xi)$$

$$= \sum_{j \in \mathbb{Z}} \int_{\mathbb{R}^n} \Delta_j f(x)\overline{\tilde{\Delta}_j g(x)} \, d\mathscr{L}(x)$$

$$= \int_{\mathbb{R}^n} \sum_{j \in \mathbb{Z}} \Delta_j f(x)\tilde{\Delta}_j g(x) \, d\mathscr{L}(x)$$

where $\tilde{\Delta}_j g = \tilde{\varphi}_j * g$ and this implies

$$\left| \int_{\mathbb{R}^n} f\overline{g} \, d\mathscr{L} \right| \leq \big\| |Tf|_{\ell^2(\mathbb{Z})} \big\|_p \big\| |\tilde{T}g|_{\ell^2(\mathbb{Z})} \big\|_{p'}.$$

A repetition of previous steps applied to $\tilde{T}$ yields $\big\| |\tilde{T}g|_{\ell^2(\mathbb{Z})} \big\|_p \leq C\|g\|_{p'}$. By duality of L^p and $L^{p'}$ and density of $C_c^\infty(\mathbb{R}^n)$ in $L^p(\mathbb{R}^n)$,

$$\|f\|_p \leq C(p', n, \varphi) \leq \big\| |Tf|_{\ell^2(\mathbb{Z})} \big\|_p.$$

Then, we can remove the $f \in L^2(\mathbb{R}^n; \mathbb{C})$ again by density and boundedness of T.

$\square$

Remark 7.6.8. *One can prove a "continuous time" Littlewood Paley estimates with same φ. Define for $t > 0$*

$$\varphi_t(x) = \frac{1}{t^n}\varphi(\frac{x}{t}) \qquad and \qquad Q_t = \varphi_t * f.$$

Then, for all $p \in (1,\infty)$, there exists constants $C_1, C_2 < \infty$ such that for all $f \in L^p(\mathbb{R}^n)$,

$$C_1\|f\|_p \leq \left\| \left(\int_0^\infty |Q_t f|^2 \, \frac{dt}{t} \right)^{1/2} \right\|_p \leq C_2\|f\|_p.$$

That is, we replaced 2^j to $\frac{1}{t}$ and $\ell^2(\mathbb{Z})$ to $L^2(\mathbb{R}_+, \frac{dt}{t})$. This is because

$$\int_0^\infty h(t) \, dt = \sum_{j\in\mathbb{Z}} \int_{2^{-j}}^{2^{-(j+1)}} h(t) \, \frac{dt}{t}.$$

This perspective tells us that $\ell^2(\mathbb{Z})$ is really the discrete version of $L^2(\mathbb{R}_+, \frac{dt}{t})$.

Remark 7.6.9. *If $f \in \mathscr{S}'(\mathbb{R}^n)$ with $(\sum_{j\in\mathbb{Z}} |\Delta_j f|^2)^{\frac{1}{2}} \in L^p(\mathbb{R}^n; \mathbb{C})$, what can we conclude? The answer is that there exists a polynomial $P \in \mathbb{C}[X_1,\ldots,X_n]$ such that $f + P \in L^p(\mathbb{R}^n; \mathbb{C})$. The same conclusion happens if one assumes instead $\left(\int_0^\infty |Q_t f|^2 \, \frac{dt}{t} \right)^{1/2} \in L^p(\mathbb{R}^n; \mathbb{C})$ in the previous remark.*

Chapter 8

Carleson measures and BMO

Definition 8.0.1 (Carleson measure). *A Carleson measure is a positive locally finite Borel measure μ on $\mathbb{R}_+^{n+1}$ such that there exists a constant $C < \infty$ for which*

$$\mu(B \times (0, r]) \le C\mathscr{L}(B)$$

for all $B = B(x, r)$. We call $B \times (0, r]$ the Carleson window and define the Carleson norm

$$\|\mu\|_{\mathscr{C}} = \sup_B \frac{\mu(B \times (0, \operatorname{rad} B])}{\mathscr{L}(B))}.$$

The following measures are *not* Carleson measures.

Example 8.0.2. *(i) $d\mu(x, t) = d\mathscr{L}(x)dt$ since no such constant C is possible for large balls.*

(ii) $d\mu(x, t) = d\mathscr{L}(x)\frac{dt}{t}$ since $\mu(B \times (0, r]) = \mathscr{L}(B) \int_0^r \frac{dt}{t} = \infty$.

(iii) $d\mu(x, t) = \frac{d\mathscr{L}(x)dt}{t^\alpha}$ for $\alpha \in \mathbb{R}$. Note that

$$\mu(B \times (0, r]) = \mathscr{L}(B) \int_0^r \frac{dt}{t^\alpha} = \begin{cases} \mathscr{L}(B)\frac{r^{1-\alpha}}{1-\alpha} & 1 - \alpha > 0 \\ \infty & otherwise \end{cases}.$$

So, we only need to consider the situation $1 - \alpha > 0$ but in this case, we cannot get uniform control via a constant C.

Definition 8.0.3 (Cone). *Let $x \in \mathbb{R}^n$. We define the cone over x:*

$$\Gamma(x) = \left\{ (y, t) \in \mathbb{R}_+^{n+1} : |x - y| < t \right\}.$$

The following are examples of Carleson measures.

Example 8.0.4. *(i) $d\mu(x, t) = \chi_{[a,b]}(t)d\mathscr{L}(x)\frac{dt}{t}$ where $0 < a < b < \infty$. Then, the constant $C = \ln\frac{b}{a}$.*

(ii) $d\mu(y,t) = \chi_{\Gamma(x)}(y)d\mathscr{L}(y)\frac{dt}{t}$. *Then,*

$$\mu(B \times (0,r]) \leq \int_0^r \mathscr{L}(B(x,t))\,\frac{dt}{t} = \int_0^r t^n \mathscr{L}(B(0,1))\,\frac{dt}{t} = \frac{\mathscr{L}(B)}{n}.$$

Definition 8.0.5 (Tent). *Let $B = B(x_B, r_B) \subset \mathbb{R}^n$ be an open ball. We define the Tent over the ball:*

$$T(B) = \left\{(y,t) \in \mathbb{R}_+^{n+1} : 0 < t \leq d(y, {}^c B)\right\} = \left\{(y,t) \in \mathbb{R}_+^{n+1} : B(y,t) \subset B\right\}.$$

(Balls are open in this definition) Similarly, using d_∞ instead of d, and $\ell(Q)$ instead of r, we can define a Tent over a cube Q.

Remark 8.0.6. *1. $B \times (0,r]$ can be changed to $B \times (0, cr]$ for any fixed $c > 0$.*

2. $B \times (0,r]$ can be changed to $T(B)$.

8.1 Geometry of Tents and Cones

We begin with the following observation.

Proposition 8.1.1. *Let B be an open ball. Then, ${}^c T(B) = \cup_{z \notin B}\Gamma(z)$.*

This lead to the following definition.

Definition 8.1.2 (Tent over open set). *Let $\Omega \subset \mathbb{R}^n$ be open. Then, define*

$$T(\Omega) = \left\{(y,t) \in \mathbb{R}_+^{n+1} : 0 < t \leq d(y, {}^c \Omega)\right\} = {}^c\left(\cup_{z \notin \Omega}\Gamma(z)\right).$$

Remark 8.1.3. *Observe that $(y,t) \in T(\Omega)$ implies that $y \in \Omega$.*

Definition 8.1.4 (Non-tangential maximal function). *Let $f : \mathbb{R}_+^{n+1} \to \mathbb{C}$ and define the non-tangential maximal function of f:*

$$\mathcal{M}^* f(x) = \sup_{(y,t) \in \Gamma(x)} |f(y,t)| \in [0, \infty].$$

Remark 8.1.5. *Given a Borel measure μ on $\mathbb{R}_+^{n+1}$, we can define the non-tangential maximal function $\mathcal{M}_\mu^*$ with respect to μ by replacing the $\sup$ with an esssup. Note then that $\mathcal{M}_\mu^*$ is defined μ almost everywhere.*

Proposition 8.1.6. *$\mathcal{M}^* f$ is lower semicontinous and hence Borel.*

Proof. Let $\alpha \geq 0$ and $x \in \mathbb{R}^n$ such that $\mathcal{M}^* f(x) > \alpha$. Now, there exists a $(y,t) \in \Gamma(x)$ such that $|f(y,t)| > \alpha$. Therefore, for all $z \in B(y,t)$ we have $(y,t) \in \Gamma(z)$ and hence $\mathcal{M}^*(z) \geq |f(y,t)| > \alpha$. That is, $x \in B(y,t) \subset \{x \in \mathbb{R}^n : \mathcal{M}^* f(x) > \alpha\}$. $\qquad\square$

Proposition 8.1.7. *Fix $\alpha \geq 0$. Then,*

$$\left\{(y,t) \in \mathbb{R}_+^{n+1} : |f(y,t)| > \alpha\right\} \subset T(\{x \in \mathbb{R}^n : \mathcal{M}^* f(x) > \alpha\}).$$

Proof. Note that $|f(y,t)| > \alpha$ implies $B(y,t) \subset \{x \in \mathbb{R}^n : \mathcal{M}^* f(x) > \alpha\}$ and hence $t < d(y, {}^c\Omega)$. $\qquad\square$

Definition 8.1.8. *Define:*

$$\mathcal{C}(\mu)(x) = \sup_{B \ni x} \frac{\mu(T(B))}{\mathscr{L}(B)} \in [0, \infty].$$

Theorem 8.1.9 (Carleson's Lemma)**.** *Let μ be a Carleson measure and $f : \mathbb{R}^{n+1}_+ \to \mathbb{C}$ be a μ-measurable function. Let $\alpha > 0$ such that $\mathscr{L}\{x \in \mathbb{R}^n : \mathcal{M}^* f(x) > \alpha\} < \infty$. Then there exists a $C(n)$ such that*

$$\mu\{(y,t) \in \mathbb{R}^{n+1}_+ : |f(y,t)| > \alpha\} \leq C(n) \int_{\{x \in \mathbb{R}^n : \mathcal{M}^* f(x) > \alpha\}} \mathcal{C}(\mu)(x) \, d\mathscr{L}(x).$$

Remark 8.1.10. *(i) For all $x \in \mathbb{R}^n$ $\mathcal{C}(\mu)(x) \leq \|\mu\|_{\mathscr{C}}$ (with $\|\cdot\|_{\mathscr{C}}$ defined using tents or Carleson windows).*

(ii) For all ball B,
$$\frac{\mu(T(B))}{\mathscr{L}(B)} \leq \inf_{x \in B} \mathcal{C}(\mu)(x).$$

(iii) $\mathcal{C}(\mu)$ is lower semicontinous and non negative.

Proof of Carleson's Lemma. Set $\Omega = \{x \in \mathbb{R}^n : \mathcal{M}^* f(x) > \alpha\}$ and note that Ω is open with $\mathscr{L}(\Omega) < \infty$. Hence, $\Omega \subsetneq \mathbb{R}^n$ and so we can apply the Whitney Covering Lemma (with balls) to obtain the existence of $c = c(n), C = C(n)$ with $c(n) < 1 < C(n) < \infty$ and $\{B_i = B_i(y_i, r_i)\}_{i \in I}$ where each B_i is a ball, $\Omega = \cup_i B_i$ and cB_i are mutually disjoint and $CB_i \cap {}^c\Omega \neq \varnothing$.

We note that it is enough to estimate $\mu(T(\Omega))$. Let $(y,t) \in T(\Omega)$. Then, $y \in \Omega$ and there exists an $i \in I$ such that $y \in B_i$. Let $z \in CB_i \cap {}^c\Omega$. Then,

$$t \leq d(y, {}^c\Omega) \leq |y - z| \leq |y - y_i| + |y_i - z| \leq (1 + C)r_i \leq d(y, {}^c(2+C)B_i).$$

Thus, $(y,t) \in T((2+C)B_i)$. Then,

$$\begin{aligned}
\mu(T(\Omega)) &\leq \mu(\cup_i T((2+C)B_i)) \\
&\leq \sum_i \mu(T((2+C)B_i)) \\
&\leq \sum_i \inf_{z \in (2+C)B_i} \mathcal{C}(\mu)(x)\mathscr{L}((2+C)B_i) \\
&\leq \sum_i \inf_{z \in cB_i} \mathcal{C}(\mu)(z) \left(\frac{2+C}{c}\right)^n \mathscr{L}(cB_i) \\
&\leq \left(\frac{2+C}{c}\right)^n \sum_i \int_{cB_i} \mathcal{C}(\mu)(z) \, d\mathscr{L}(z) \\
&\leq \left(\frac{2+C}{c}\right)^n \int_{\Omega} \mathcal{C}(\mu)(z) \, d\mathscr{L}(z)
\end{aligned}$$

since the balls cB_i are mutually disjoint in Ω. $\qquad\square$

Corollary 8.1.11. *For all open* Ω, $\mu(T(\Omega)) \leq C\|\mu\|_{\mathscr{C}}\mathscr{L}(\Omega)$.

Proof. Note that in the previous proof, we only used the fact that $\Omega = \{x \in \mathbb{R}^n : \mathcal{M}^* f(x) > \alpha\}$ to obtain that $\Omega \subsetneq \mathbb{R}^n$. So, the argument works when $\Omega \subsetneq \mathbb{R}^n$. Certainly, the claim is trivially true when $\Omega = \mathbb{R}^n$. $\square$

Corollary 8.1.12. *With the assumptions of Carleson's Lemma,*

$$\mu\left\{(y,t) \in \mathbb{R}^{n+1}_+ : |f(y,t)| > \alpha\right\} \leq c\|\mu\|_{\mathscr{C}}\mathscr{L}(\{x \in \mathbb{R}^n : \mathcal{M}^* f(x) > \alpha\}).$$

Corollary 8.1.13. *Assume* $f : \mathbb{R}^{n+1}_+ \to \mathbb{C}$ *is* μ-*measurable and* μ *is a Carleson measure. Then,*

$$\iint_{\mathbb{R}^{n+1}_+} |f(y,t)| \, d\mu(y,t) \leq C\|\mu\|_{\mathscr{C}} \int_{\mathbb{R}^n} \mathcal{M}^* f(x) \, d\mathscr{L}(x).$$

Proof. If $\|\mathcal{M}^* f\|_1 = \infty$, there's nothing to prove. So, assume not. Then, $\|\mathcal{M}^* f\|_1 < \infty$ implies $\mathscr{L}\{x \in \mathbb{R}^n : \mathcal{M}^* f(x) > \alpha\} < \infty$ for all $\alpha > 0$. Thus,

$$\mu\left\{(y,t) \in \mathbb{R}^{n+1}_+ : |f(y,t)| > \alpha\right\} \leq C(n)\mathscr{L}\{x \in \mathbb{R}^n : \mathcal{M}^* f(x) > \alpha\}\|\mu\|_{\mathscr{C}}$$

and integrating both sides from 0 to ∞ in α finishes the proof. $\square$

Definition 8.1.14. *Fix* $t > 0$ *and let* φ *be a function on* $\mathbb{R}^n$. *We define*

$$\varphi_t(x) = \frac{1}{t^n}\varphi\left(\frac{x}{t}\right).$$

We state the following technical Lemma.

Lemma 8.1.15. *Let* $\varphi(x) = (1 + |x|)^{-n-\varepsilon}$, $\varepsilon > 0$, $x \in \mathbb{R}^n$. *Then, there exists a* $c = c(n,\varepsilon) < \infty$ *such that for all* $x \in \mathbb{R}^n$ *and for all* $(y,t) \in \Gamma(x)$

$$|\varphi_t * f(y)| \leq c\mathcal{M}f(x)$$

for all $f \in \mathrm{L}^1_{\mathrm{loc}}(\mathbb{R}^n)$ *such that* $\int_{\mathbb{R}^n} \varphi(y)|f(y)| \, d\mathscr{L}(y) < \infty$.

We define the following operator family which will be of interest to us.

Definition 8.1.16 $((R_t)_{t>0})$. *For* $t > 0$, *let*

$$R_t f(x) = \int_{\mathbb{R}^n} K_t(x,y)f(y) \, d\mathscr{L}(y)$$

almost everywhere $x \in \mathbb{R}^n$ *whenever* $f \in \mathrm{L}^1_{\mathrm{loc}}(\mathbb{R}^n)$ *such that* $\int_{\mathbb{R}^n} \varphi(y)|f(y)| \, d\mathscr{L}(y) < \infty$ *and with*

$$|K_t(x,y)| \leq C\varphi_t(x-y)$$

for all $t > 0$ *and almost everywhere* $x, y \in \mathbb{R}^n$.

Corollary 8.1.17 (Carleson Embedding). *For all* $1 < p < \infty$ *and* $f \in \mathrm{L}^p(\mathbb{R}^n)$,

$$\iint_{\mathbb{R}^{n+1}_+} |R_t f(y)|^p \, d\mu(y,t) \leq C(n,\varepsilon,C)\|\mu\|_{\mathscr{C}}\|f\|_p^p$$

whenever μ *is a Carleson measure.*

Proof. Let $g(x,t) = |R_t f(x)|^p$. By application of Corollary 8.1.13,

$$\iint_{\mathbb{R}^{n+1}_+} |R_t f(x)|^p \, d\mu(x,t) \le C(n) \|\mu\|_{\mathscr{C}} \int_{\mathbb{R}^n} \mathcal{M}^* y(x) \, d\mathscr{L}(x).$$

Then, by Lemma 8.1.15, $|R_t f(y)| \le C(n,\varepsilon)\mathcal{M}f(x)$ if $|x - y| < t$. Thus,

$$C(n)\|\mu\|_{\mathscr{C}} \int_{\mathbb{R}^n} \mathcal{M}^* g(x) \, d\mathscr{L}(x) \le C(n)C(n,\varepsilon)^p \int_{\mathbb{R}^n} (\mathcal{M}f)^p \, d\mathscr{L} \le C(p,n)\|f\|_p^p.$$

$\square$

8.2 BMO and Carleson measures

Theorem 8.2.1. *Suppose that*

(i) $R_t(1)(x) = 0$ *almost everywhere* $x \in \mathbb{R}^n$ *and for all* $t > 0$ *(note that this is saying* $\int_{\mathbb{R}^n} K_t(x,y) \, d\mathscr{L}(y) = 0$ *almost everywhere),*

(ii) *For all* $f \in \mathrm{L}^2(\mathbb{R}^n)$,

$$\iint_{\mathbb{R}^{n+1}_+} |R_t f(x)|^2 \, \frac{d\mathscr{L}(x)dt}{t} \le C_1 \int_{\mathbb{R}^n} |f(x)|^2 \, d\mathscr{L}(x).$$

Let $b \in \mathrm{BMO}$. *Then* $R_t(b)(x)$ *is defined almost everywhere* $x \in \mathbb{R}^n$ *and for all* $t > 0$ *and there exists a* $C_2 < \infty$ *such that* $|R_t(b)(x)|^2 \le C_2\|b\|_*^2$ *and*

$$|R_t(b)(x)|^2 \frac{d\mathscr{L}(x)dt}{t}$$

is a Carleson measure.

Remark 8.2.2. *The estimate in condition (ii) of the Theorem is one of Littlewood-Paley type. It does* not *follow from the kernel bounds on* R_t. *That is,* $\|R_t f\|_2 \le C(n,\varepsilon)\|f\|_2$ *is not enough to integrate from 0 to ∞ in* $\frac{dt}{t}$.

Proof of Theorem 8.2.1. Fix $x \in \mathbb{R}^n$. Pick a ball B centred at x. Write $b = b_1 + b_2 + \mathrm{m}_B b$ where $b_1 = (b - \mathrm{m}_B b)\chi_B$ and $b_2 = (b - \mathrm{m}_B b)\chi_{{}^c B}$. Let

$$I_j = \int_{\mathbb{R}^n} |K_t(x,y)| \, |b_j(y)| \, d\mathscr{L}(y)$$

for $j = 1, 2$. We estimate these two quantities.

First, assume that $\mathrm{rad}\, B = t$. Then,

$$I_1 \le \int_B \frac{1}{t^n} \varphi\left(\frac{x - y}{t}\right) |b(y) - \mathrm{m}_B b| \, d\mathscr{L}(y) \le C(n)\|b\|_*$$

and using the fact that when $|x - y| \sim 2^n t$,

$$\frac{1}{t^n} \varphi \left(\frac{x - y}{t} \right) \leq \frac{1}{t^n} \left(1 + \frac{x - y}{t} \right)^{-n - \varepsilon} \leq \frac{1}{t^n} 2^{-j(n + \varepsilon)}$$

we have

$$
\begin{aligned}
I_2 &\leq \int_{{}^c B} \frac{1}{t^n} \varphi \left(\frac{x - y}{t} \right) |b(y) - \mathrm{m}_B b| \; d\mathscr{L}(y) \\
&\leq \sum_j \int_{2^{j+1} B \backslash 2^j B} \frac{1}{t^n} \varphi \left(\frac{x - y}{t} \right) \left(|b - \mathrm{m}_{2^{j+1} B} b| + |\mathrm{m}_{2^{j+1} B} b - \mathrm{m}_B b| \right) \; d\mathscr{L}(y) \\
&\leq \sum_j 2^{-j\varepsilon} \frac{1}{(2^j t)^n} \left[\int_{2^{j+1} B} |b - \mathrm{m}_{2^{j+1} B} b| \; d\mathscr{L} + \int_{2^{j+1} B} |\mathrm{m}_{2^{j+1} B} b - \mathrm{m}_B b| \; d\mathscr{L} \right] \\
&\leq \sum_j 2^{-j\varepsilon} \left[C(n) \|b\|_* + C(n)(1 + j) \|b\|_* \right] \\
&\leq \sum_j (1 + j) 2^{-j\varepsilon} C(n) \|b\|_*.
\end{aligned}
$$

Therefore, $R_t b_1, R_t b_2$ are well defined with uniform estimates and $R_t(\mathrm{m}_B b) = 0$ since $\mathrm{m}_B b$ is constant by (i). Hence, $R_t b$ is well defined and $|R_t b| \leq C_2 \|b\|_*$. In fact, $R_t b(x)$ does not depend on the choices of $B \ni x$. That is, it is independent of decomposition.

Now, we do the Carleson measure estimate. Fix $B_0 = B_0(x_0, r)$ and $B = B(x_0, 2r)$. We use (ii),

$$
\begin{aligned}
\iint_{B_0 \times (0, r]} |R_t(b_1)(x)|^2 \; \frac{d\mathscr{L}(x) dt}{t} &\leq \iint_{\mathbb{R}^{n+1}_+} |R_t(b_1)(x)|^2 \; \frac{d\mathscr{L}(x) dt}{t} \\
&\leq C_1 \int_{\mathbb{R}^n} |b_1(x)|^2 \; d\mathscr{L}(x) \\
&\leq C_1 \mathscr{L}(B) \|b\|_*^2
\end{aligned}
$$

and $\mathscr{L}(B) = 2^n \mathscr{L}(B_0)$.

For $R_t(b_2)$, we do the same as before for I_2, but this time with $x \in B_0$, $y \notin B$ and $\mathrm{rad}\, B = 2r_0$. So, noting that when $|x - y| > 2^{j+1} r_0$ we have

$$
\begin{aligned}
\frac{1}{t^n} \varphi \left(\frac{x - y}{t} \right) &\leq \frac{1}{t^n} \left(1 + \frac{|x - y|}{t} \right)^{-n - \varepsilon} \\
&\leq \frac{1}{t^n} \frac{t^{n + \varepsilon}}{(2^{j+1} r_0)^{n + \varepsilon}} = \left(\frac{t}{r_0} \right)^{\varepsilon} \left(\frac{1}{2^{j+1} r_0} \right)^n 2^{-(j+1)\varepsilon}
\end{aligned}
$$

we calculate

$$I_2 \leq \int_{cB} \frac{1}{t^n} \varphi\left(\frac{x-y}{t}\right) |b(y) - \mathrm{m}_B\, b|\ d\mathscr{L}(y)$$

$$\leq \sum_j \int_{2^{j+1}B \setminus 2^j B} \frac{1}{t^n} \varphi\left(\frac{x-y}{t}\right) \left(|b(y) - \mathrm{m}_{2^{j+1}B}\, b| + |\mathrm{m}_{2^{j+1}B} - \mathrm{m}_B\, b|\right)\ d\mathscr{L}(y)$$

$$\leq \sum_j \left(\frac{t}{r_0}\right)^\varepsilon 2^{-(j+1)\varepsilon} \left(\frac{1}{2^{j+1}r_0}\right)^n \int_{2^{j+1}B} \left(|b(y) - \mathrm{m}_{2^{j+1}B}\, b| + |\mathrm{m}_{2^{j+1}B} - \mathrm{m}_B\, b|\right)\ d\mathscr{L}(y)$$

$$\leq \left(\frac{t}{r_0}\right)^\varepsilon \sum_j (1+j) 2^{-(j+1)\varepsilon} C(n)\|b\|_*.$$

Thus,

$$\iint_{B_0 \times (0,r]} |R_t(b_2)(x)|^2\ \frac{d\mathscr{L}(x)dt}{t} \leq C(n,\varepsilon)\|b\|_*^2 \iint_{B_0 \times (0,r]} \left(\frac{t}{r_0}\right)^{2\varepsilon} \frac{d\mathscr{L}(x)dt}{t}$$

$$\leq C(n,\varepsilon)\|b\|_*^2 \frac{1}{2\varepsilon}\mathscr{L}(B_0).$$

the proof is completed by adding up the two estimates. $\square$

We define the following family of operators.

Definition 8.2.3 $((Q_t)_{t>0})$. *Let* $\varphi \in \mathscr{S}(\mathbb{R}^n)$ *with* $\int_{\mathbb{R}^n} \varphi\ d\mathscr{L} = 0$. *Define* $Q_t f = \varphi_t * f$ *when* $f \in \mathrm{L}^1_{\mathrm{loc}}(\mathbb{R}^n)$.

Remark 8.2.4. *We note that*

$$Q_t f(x) = \int_{\mathbb{R}^n} \frac{1}{t^n} \varphi\left(\frac{x-y}{t}\right) f(y)\ d\mathscr{L}(y).$$

We have (i):

$$Q_t(1) = \int_{\mathbb{R}^n} \varphi_t(x-y)\ d\mathscr{L}(y) = 0.$$

And for (ii):

$$\iint_{\mathbb{R}^{n+1}_+} |Q_t f(x)|^2\ \frac{d\mathscr{L}(x)dt}{t} = \frac{1}{(2\pi)^n} \int_{\mathbb{R}^n} \left(\int_0^\infty |\hat{\varphi}(t\xi)|^2\right) \frac{dt}{t} \left|\hat{f}(\xi)\right|^2\ d\mathscr{L}(\xi)$$

$$\leq A\|f\|_2^2.$$

since $\hat{\varphi}(0) = 0$ *and*

$$|\hat{\varphi}(\xi)| \leq \begin{cases} c|\xi| & |\xi| \leq 1 \\ \frac{c}{|\xi|} & |\xi| \geq 1 \end{cases}$$

with

$$A = \sup_{\xi \in \mathbb{R}^n} \int_0^\infty |\hat{\varphi}(t\xi)|\ \frac{dt}{t} < \infty.$$

So, we can apply the Theorem with $R_t = Q_t$.

Theorem 8.2.5. *Let $\beta \in \mathrm{L}^\infty(\mathbb{R}_+^{n+1}, \frac{d\mathscr{L}(x)dt}{t})$ and assume that*

$$d\mu(x,t) = |\beta(x,t)|^2 \frac{d\mathscr{L}(x)dt}{t}$$

is a Carleson measure. Let $\beta_t(x) = \beta(x,t)$. Then for all $f \in \mathrm{H}^1(\mathbb{R}^n)$,

(i) For all $\varepsilon > 0$ and $R < \infty$,

$$I_{\varepsilon,R}(f) = \int_{\mathbb{R}^n} \int_\varepsilon^R Q_t(\beta_t)(x) f(x) \, \frac{d\mathscr{L}(x)dt}{t}$$

is well defined and $|I_{\varepsilon,R}(f)| \leq C(n,\varphi)\|f\|_{\mathrm{H}^1}\|\mu\|_{\mathscr{C}}^{\frac{1}{2}}$.

(ii) We have that

$$\lim_{\varepsilon \to 0, R \to \infty} I_{\varepsilon,R}(f)$$

exists and defines an element $b \in \mathrm{BMO}$ with $\|b\|_* \leq C(n,\varphi)\|\mu\|_{\mathscr{C}}^{\frac{1}{2}}$. We write,

$$b = \int_0^\infty Q_t\beta_t \, \frac{dt}{t}.$$

Proof. To prove (i), we note that $|Q_t(\beta_t)(x)| \leq \|\beta\|_\infty\|\varphi\|_1$ almost everywhere $(x,t) \in \mathbb{R}_+^{n+1}$. Now, $f \in \mathrm{H}^1(\mathbb{R}^n)$ implies $f \in \mathrm{L}^1(\mathbb{R}^n)$ and

$$I_{\varepsilon,R}(f) = \int_{\mathbb{R}^n} \int_\varepsilon^R Q_t(\beta_t)(x) f(x) \, \frac{d\mathscr{L}(x)dt}{t} \leq \|\beta\|_\infty\|\varphi\|_1\|f\|_1 \int_\varepsilon^R \frac{dt}{t} < \infty.$$

Now for (ii), let $f = a \in \mathscr{A}^\infty$ with spt $a \subset B$ with B a ball. Then,

$$I_{\varepsilon,R}(a) = \int_\varepsilon^R \left(\int_{\mathbb{R}^n} \beta_t(x) Q_t^{\mathrm{tr}}(a)(x) \, d\mathscr{L}(x) \right) \frac{dt}{t}$$

where $Q_t^{\mathrm{tr}} = \psi_t *$ where $\psi(y) = \varphi(-y)$. Let

$$I = \iint_{\mathbb{R}_+^{n+1}} |\beta_t(x)| \, |Q_t^{\mathrm{tr}}(a)(x)| \, \frac{d\mathscr{L}(x)dt}{t}.$$

Then clearly, $|I_{\varepsilon,R}| \leq I$.

We compute this integral by covering $\mathbb{R}_+^{n+1}$ with square annuli. Let $A_{2^jB} = 2^jB \times (0, 2^jr]$, $C_0 = A_{2B}$ and $C_j = A_{2^{j+1}B} \setminus A_{2^jB}$ when $j > 0$ Also, define $C_j^2 = A_{2^{j+1}B} \cap 2^jB \times (0, 2^{j+1}r]$ when $j > 0$ and $C_j^1 = C_j \setminus C_j^2$.

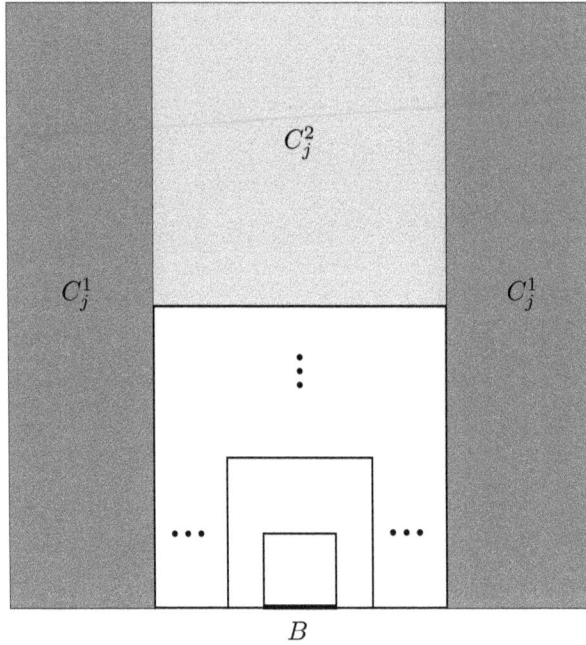

For $j = 0$, by application of Cauchy-Schwarz,

$$\iint_{C_0} |\beta_t(x)| \, |Q_t^{\mathrm{tr}}(a)(x)| \, \frac{d\mathscr{L}(x)dt}{t} \leq \left(\iint_{C_0} |\beta_t(x)|^2 \, \frac{d\mathscr{L}(x)dt}{t} \right)^{\frac{1}{2}} \left(\iint_{C_0} |Q_t^{\mathrm{tr}}(a)(x)|^2 \, \frac{d\mathscr{L}(x)dt}{t} \right)^{\frac{1}{2}}$$

$$\leq \|\mu\|_{\mathscr{C}} \mathscr{L}(2B)^{\frac{1}{2}} C(n, \varphi) \|a\|_2.$$

But $\|a\|_2 \leq C(n, \varphi) \mathscr{L}(B)^{-\frac{1}{2}}$.

Now, we consider $j > 0$. Let

$$A(x, t) = \sup_{y \in B} \frac{1}{t^n} \left| \varphi \left(\frac{y - x}{t} \right) - \varphi \left(\frac{y_B - x}{t} \right) \right|$$

so that we have $\left| Q_t^{\mathrm{tr}}(a)(x) \right| \leq A(x, t) \|a\|_1$. Consider $(x, t) \in C_j^1$. Then, $2^j \leq |x - y_B| \leq 2^{j+1}r$ and $0 < t \leq 2^{j+1}r$. If $y \in B$ then $|x - y| \sim 2^j r$ and we have the mean value inequality

$$|A(x, t)| \leq C(\varphi, n) \frac{r}{t} \frac{1}{t^n} \left(1 + \frac{2^j}{t} \right)^{-M}$$

for all $M > 0$. Therefore,

$$\iint_{C_j^1} |A(x, t)|^2 \, \frac{d\mathscr{L}(x)dt}{t} \leq \mathscr{L}(2^{j+1}B \setminus 2^j B) \int_0^{2^{j+1}r} C \left(\frac{r}{t} \frac{1}{t^n} \left(1 + \frac{2^j}{t} \right)^{-M} \right)^2 \frac{dt}{t}$$

$$\leq C(2^{j+1}r)^n \frac{1}{2^{2j}} \frac{1}{(2^j r)^{2n}} \int_1^\infty (u^{n+1}(1 + u)^{-M})^2 \frac{du}{u}$$

$$\leq C \frac{1}{2^{2j}} \frac{1}{\mathscr{L}(2^{j+1}B)}$$

since $\int_1^\infty (u^{n+1}(1 + u)^{-M})^2 \frac{du}{u} < \infty$ when $M > n + 1$.

89

Now, if $(x, t) \in C_j^2$, we have $|x - y_B| \leq 2^j r$, and $2^j r \leq t < 2^{j+1} r$ and the mean value inequality takes the form

$$A(x, t) \leq \sup_{z \in B} \left(\frac{r}{t} \frac{1}{t^n} \left| \nabla \varphi \left(\frac{z - x}{t} \right) \right| \right)$$

and when $x \in 2B$,

$$\left(\frac{r}{t} \frac{1}{t^n} \left| \nabla \varphi \left(\frac{z - x}{t} \right) \right| \right) \leq \frac{r}{(2^j r)^{n+1}} \| \nabla \varphi \|_\infty$$

and when $x \notin 2B$,

$$A(x, t) \leq C(n) \frac{r}{(2^j r)^{n+1}} \left(1 + \frac{|y_B - x|}{2^j r} \right)^{-M}.$$

Therefore,

$$\iint_{C_j^2} |A(x, t)|^2 \, \frac{d\mathscr{L}(x) dt}{t} \leq \int_{2_j B} \int_{2^j r}^{2^{j+1} r} |A(x, t)|^2 \, \frac{d\mathscr{L}(x) dt}{t}$$

$$\leq \ln 2 \int_{2^j B} |A(x, t)|^2 \, d\mathscr{L}(x)$$

$$\leq \ln 2 \left(\int_{2B} |A(x, t)|^2 \, d\mathscr{L}(x) + \int_{2^j B \backslash 2B} |A(x, t)|^2 \, d\mathscr{L}(x) \right)$$

$$\leq C \left(\frac{1}{2^{j(n+2)} \mathscr{L}(B)} + \frac{r^2}{(2^j r)^{2(n+1)}} \int_{2^j B} \left(1 + \frac{|y_B - x|}{2^j r} \right)^{-2M} \, d\mathscr{L}(x) \right)$$

$$\leq C \left(\frac{1}{2^{2j} \mathscr{L}(B)} + \frac{1}{2^{2j} \mathscr{L}(2^j B)} \right)$$

Therefore,

$$\iint_{C_j^2} |A(x, t)|^2 \, \frac{d\mathscr{L}(x) dt}{t} \leq \frac{C}{2^{2j}} \frac{1}{\mathscr{L}(2^{j+1} B)}$$

and

$$I = \iint_{\mathbb{R}_+^{n+1}} |\beta_t(x)| \, |Q_t^{\mathrm{tr}}(a)(x)| \, \frac{d\mathscr{L}(x) dt}{t}$$

$$= \sum_j \iint_{C_j} |\beta_t(x)| \, |Q_t^{\mathrm{tr}}(a)(x)| \, \frac{d\mathscr{L}(x) dt}{t}$$

$$\leq \sum_j \left(\iint_{C_j} |\beta_t(x)|^2 \, \frac{d\mathscr{L}(x) dt}{t} \right)^{\frac{1}{2}} \left(\iint_{C_j} |Q_t^{\mathrm{tr}}(a)(x)|^2 \, \frac{d\mathscr{L}(x) dt}{t} \right)^{\frac{1}{2}}$$

$$\leq (\|\mu\|_{\mathscr{C}} \mathscr{L}(2^{j+1} B))^{\frac{1}{2}} \left(\frac{C}{2^{2j}} \frac{1}{\mathscr{L}(2^{j+1} B)} \right)^{\frac{1}{2}}$$

$$\leq C \|\mu\|_{\mathscr{C}}^{\frac{1}{2}}.$$

This tells us that $\lim_{\varepsilon \to 0, \ R \to \infty} I_{\varepsilon, R}$ exists and

$$\lim_{\varepsilon \to 0, \ R \to \infty} I_{\varepsilon, R} = \iint_{\mathbb{R}_+^{n+1}} \beta_t(x) Q_t^{\mathrm{tr}}(a)(x) \, \frac{d\mathscr{L}(x) dt}{t}.$$

Now, take a general $f \in \mathrm{H}^1$. Then, we can find λ_j and $a_j \in \mathscr{A}^\infty$ such that $f = \sum_j \lambda_j a_j$ in H^1 and in particular $\mathrm{L}^1(\mathbb{R}^n)$ and also such that $\sum_j |\lambda_j| \le 2\|f\|_{\mathrm{H}^1}$. Using the fact that $I_{\varepsilon,R} \le C_{\varepsilon,R}\|f\|_1$, we can write

$$I_{\varepsilon,R}(f) = \sum_j \lambda_j I_{\varepsilon,R}(a_j).$$

Therefore,

$$|I_{\varepsilon,R}(f)| \le \sum_j |\lambda_j|\, |I_{\varepsilon,R}(a_j)| \le C\|\mu\|_{\mathscr{C}}^{\frac{1}{2}} \sum_j |\lambda_j| \le C\|\mu\|_{\mathscr{C}}^{\frac{1}{2}} 2\|f\|_{\mathrm{H}^1}$$

with C independent of ε and R.

Let

$$b_{\varepsilon,R} = \int_\varepsilon^R Q_t \beta_t \, \frac{dt}{t}$$

and note that it is a measurable function. Also, $\langle b_{\varepsilon,R}, f \rangle = I_{\varepsilon,R}(f)$. Thus, $b_{\varepsilon,R} \in \mathrm{BMO}$ with $\|b_{\varepsilon,R}\|_* \le C\|\mu\|_{\mathscr{C}}^{\frac{1}{2}}$ with C independent of ε, R. Now since $\mathrm{BMO} = \mathrm{H}^{1'}$ there exists a convergent subsequence (b_{ε_j, R_j}) and $\langle b_{\varepsilon_j, R_j}, f \rangle \to \langle b, f \rangle$ for some $b \in \mathrm{BMO}$. The uniqueness of β then follows from the uniqueness of $\lim_{\varepsilon \to 0,\ R \to \infty} I_{\varepsilon,R}(a)$ for atoms $a \in \mathscr{A}^\infty$ and then by the density of $\mathrm{Vect}\, \mathscr{A}^\infty$ in H^1 so $\lim_{\varepsilon \to 0,\ R \to \infty} \langle b_{\varepsilon,R}, f \rangle = \langle b, f \rangle$. $\qquad\square$

Exercise 8.2.6. *Replace Q_t^{tr} by R_t with kernel K_t satisfying:*

$$K_t(x,y) \le \frac{C}{t^n}\left(1 + \frac{|x-y|}{t}\right)^{-n-\varepsilon}$$

when $\varepsilon > 0$ and

$$|K_t(x,y) - K_t(x,z)| \le C\left(\frac{|y-z|}{t}\right)^\delta \frac{1}{t^n}\left(1 + \frac{|x-y|}{t}\right)^{-n-\varepsilon-\delta}$$

for $\delta > 0$ when $|y-z| \le \frac{1}{2}(t + |x-y|)$.

Theorem 8.2.7 (Paraproducts of J.M. Bony). *Let $\psi, \tilde{\psi}, \varphi \in \mathscr{S}(\mathbb{R}^n)$ such that $\int_{\mathbb{R}^n} \psi \, d\mathscr{L} = \int_{\mathbb{R}^n} \tilde{\psi} \, d\mathscr{L} = 0$. Define,*

$$Q_t = \psi_t *, \quad \tilde{Q}_t = \tilde{\psi}_t *, \quad P_t = \varphi_t *$$

and let $b \in \mathrm{BMO}$. Let

$$\pi_b(f) = \int_0^\infty \tilde{Q}_t(Q_t(b)P_t(f)) \frac{dt}{t}.$$

Then,

   (i) *We have $\pi_b \in \mathcal{L}(\mathrm{L}^2(\mathbb{R}^n))$ and*

$$\|\pi_b(f)\|_2 \le C(n, \varphi, \psi, \tilde{\psi})\|b\|_*\|f\|_2$$

    whenever $f \in \mathrm{L}^2(\mathbb{R}^n)$,

  (ii) *$\pi_b \in \mathrm{CZO}_1$,*

(iii) $\pi_b^{\mathrm{tr}}(1) = 0$ *in* BMO,

(iv) If $\int_{\mathbb{R}^n} \varphi \, d\mathscr{L} = 1$ *and* $\psi, \tilde{\psi}$ *are radial functions satisfying*

$$\int_0^\infty \tilde{\psi}(t\xi)\psi(t\xi) \, \frac{dt}{t} = 1$$

then $\pi_b(1) = b$ *in* BMO.

Proof.   (i) Let

$$I_{\varepsilon,R} = \langle \int_\varepsilon^R \tilde{Q}_t(Q_t(b)P_t(f)) \, \frac{dt}{t}, g \rangle$$

whenever $g \in \mathrm{L}^2(\mathbb{R}^n)$. Then, by noting that $\langle \tilde{Q}_t(Q_t(b)P_t(f)), g \rangle = \langle Q_t(b)P_t(f), \tilde{Q}_t^{\mathrm{tr}}(g) \rangle$ and applying Fubini,

$$\begin{aligned}
|I_{\varepsilon,R}| &= \left| \int_\varepsilon^R \int_{\mathbb{R}^n} Q_t(b)(x)P_t(f)(x)\tilde{Q}_t^{\mathrm{tr}}(g)(x) \, \frac{d\mathscr{L}(x)dt}{t} \right| \\
&\leq \int_\varepsilon^R \int_{\mathbb{R}^n} \left| Q_t(b)(x)P_t(f)(x)\tilde{Q}_t^{\mathrm{tr}}(g)(x) \right| \, \frac{d\mathscr{L}(x)dt}{t} \\
&\leq I_1 \, I_2
\end{aligned}$$

where

$$I_1^2 = \iint_{\mathbb{R}_+^{n+1}} |P_t(f)(x)Q_t(b)(x)|^2 \, \frac{d\mathscr{L}(x)dt}{t}$$

and

$$I_2^2 = \iint_{\mathbb{R}_+^{n+1}} \left| \tilde{Q}_t^{\mathrm{tr}}(g)(x) \right|^2 \, \frac{d\mathscr{L}(x)dt}{t}.$$

We have the Littlewood-Paley estimate $I_2 \leq C(\tilde{\psi}, n)\|g\|_2$ since $\tilde{\psi} \in \mathscr{S}(\mathbb{R}^n)$ and $\int_{\mathbb{R}^n} \psi \, d\mathscr{L} = 0$. Then, note that

$$|Q_t(b)(x)|^2 \, \frac{d\mathscr{L}(x)dt}{t}$$

is a Carleson measure and so by application of Carleson Embedding Theorem with $R_t = P_t$,

$$I_1 \leq C(n, \varphi, \psi)\|f\|_2\|b\|_*.$$

Hence,

$$\lim_{\varepsilon \to 0, \, R \to \infty} I_{\varepsilon,R} = \iint_{\mathbb{R}_+^{n+1}} Q_t(b)(x)P_t(f)(x)\tilde{Q}_t^{\mathrm{tr}}(g)(x) \, \frac{d\mathscr{L}(x)dt}{t}$$

is a bounded bilinear form on $\mathrm{L}^2(\mathbb{R}^n)$ and we conclude (i) by invoking the Riesz Representation Theorem.

(ii) Let $f, g \in \mathrm{C}_c^\infty(\mathbb{R}^n)$. Then,

$$\langle \pi_b(f), g \rangle = \iint_{\mathbb{R}_+^{n+1}} Q_t(b)(x)P_t(f)(x)\tilde{Q}_t^{\mathrm{tr}}(g)(x) \, \frac{d\mathscr{L}(x)dt}{t} = \langle K, g \otimes f \rangle$$

where $g \otimes f(y, z) = g(y)f(z)$ and

$$K(y, z) = \int_0^\infty \int_{\mathbb{R}^n} \varphi_t(z - x)Q_t(b)(x)\tilde{\psi}_t(y - x) \, \frac{d\mathscr{L}(x)dt}{t}$$

where this equality is in $\mathscr{S}(\mathbb{R}^n)'$.

Now by using the fact that $|Q_t(b)(x)| \leq C\|b\|_*$ and by the decay of $\varphi, \tilde{\psi}$ using that $\eta_\varepsilon(x) \leq (1 + |x|)^{-n-\varepsilon}$ and $\eta_\varepsilon * \eta_\varepsilon \leq C(n, \varepsilon)\eta_\varepsilon$, we have

$$\int_0^\infty \frac{1}{t^n}\eta_\varepsilon\left(\frac{y-z}{t}\right) \frac{dt}{t} \leq \frac{C(n, \varepsilon)}{|y-z|^n}.$$

The estimate for $\nabla_y K$ follows since under differentiation in y, $\tilde{\psi}_t$ gives an extra $\frac{1}{t}$ factor and so we get

$$|\nabla_y K| \leq \frac{C(n, \varepsilon)}{|y-z|^{n+1}}.$$

So, $K \in \mathrm{CZK}_1$.

(iv) We already have that

$$\langle \pi_b(f), g \rangle = \iint_{\mathbb{R}_+^{n+1}} Q_t(b)(x)P_t(f)(x)\tilde{Q}_t^{\mathrm{tr}}(g)(x) \frac{d\mathscr{L}(x)dt}{t}$$

whenever $f, g \in \mathrm{L}^2(\mathbb{R}^n)$. First, we show this same equality when $f \in \mathrm{L}^\infty(\mathbb{R}^n)$ and when $g = a \in \mathscr{A}^\infty$.

Recall that when $b \in \mathrm{BMO}$ and $a \in \mathscr{A}^\infty$, we have

$$\iint_{\mathbb{R}_+^{n+1}} |Q_t(b)(x)|\left|\tilde{Q}_t^{\mathrm{tr}}(a)(x)\right| \frac{d\mathscr{L}(x)dt}{t} \leq C\|b\|_*.$$

Also, $|P_t(f)(x)| \leq \|\varphi\|_1\|f\|_\infty$ for all $(x, t) \in \mathbb{R}_+^{n+1}$. Therefore,

$$\iint_{\mathbb{R}_+^{n+1}} Q_t(b)(x)P_t(f)(x)\tilde{Q}_t^{\mathrm{tr}}(a)(x) \frac{d\mathscr{L}(x)dt}{t}$$

exists.

Note that $\pi_b^{\mathrm{tr}} : \mathrm{H}^1 \to \mathrm{L}^1(\mathbb{R}^n)$ and so $\pi_b^{\mathrm{tr}}(a) \in \mathrm{L}^1(\mathbb{R}^n)$ and hence $\langle f, \pi_b^{\mathrm{tr}}(a) \rangle$ well defined and by definition $\langle f, \pi_b^{\mathrm{tr}}(a) \rangle = \langle \pi_b(f), a \rangle$. Let $f_k = f\chi_{[-2^k, 2^k]^n}$ when $k \in \mathbb{N}$ and so $f_k \in \mathrm{L}^2 \cap \mathrm{L}^\infty(\mathbb{R}^n)$. So by Dominated convergence $\langle f_k, \pi_b^{\mathrm{tr}}(a) \rangle \to \langle f, \pi_b^{\mathrm{tr}}(a) \rangle$ and $P_t(f_k)(x) \to P_t(f)(x)$ for all $(x, t) \in \mathbb{R}_+^{n+1}$.

Now, set $f = 1$, and since we assumed that $\int_{\mathbb{R}^n} \varphi \, d\mathscr{L} = 1$, we have $P_t(1) = 1$ for all $(x, t) \in \mathbb{R}_+^{n+1}$. Then,

$$\langle \pi_b(1), a \rangle = \iint_{\mathbb{R}_+^{n+1}} Q_t(b)(x)\tilde{Q}_t^{\mathrm{tr}}(a) \frac{d\mathscr{L}(x)dt}{t}$$

$$= \lim_{\varepsilon \to 0, \ R \to \infty} \int_\varepsilon^R \int_{\mathbb{R}^n} Q_t(b)(x)\tilde{Q}_t^{\mathrm{tr}}(a) \frac{d\mathscr{L}(x)dt}{t} = \lim_{\varepsilon \to 0, \ R \to \infty} \langle b, U_{\varepsilon, R} \rangle$$

where

$$U_{\varepsilon, R} = \int_\varepsilon^R Q_t^{\mathrm{tr}}\tilde{Q}_t^{\mathrm{tr}}(a) \frac{dt}{t}.$$

It suffices to prove that $U_{\varepsilon, R} \to a$ in H^1 since this gives $\langle \pi_b(1), a \rangle = \langle b, a \rangle$ for all $a \in \mathscr{A}^\infty$ and hence $\pi_b(1) = b$.

We note that $a \in \mathrm{L}^2(\mathbb{R}^n)$ since spt $a \subset B$ for some ball B and $\|a\|_\infty \leq \mathscr{L}(B)^{-1}$. Using the assumption that $\psi, \tilde\psi$ are radial, we have

$$\widehat{U_{\varepsilon,R}(a)}(\xi) = \int_\varepsilon^R \hat\psi(t\xi)\hat{\tilde\psi}(t\xi)\, \frac{dt}{t} \hat a(\xi)$$

and so $U_{\varepsilon,R}(a) \to a$ in $\mathrm{L}^2(\mathbb{R}^n)$.

Define a function φ on $\mathbb{R}^n$ by

$$\varphi(\xi) = \int_1^\infty \hat\psi(t\xi)\hat{\tilde\psi}(t\xi)\, \frac{dt}{t} = 1 - \int_0^1 \hat\psi(t\xi)\hat{\tilde\psi}(t\xi)\, \frac{dt}{t}$$

whenever $\xi \in \mathbb{R}^n \setminus \{0\}$. It is an easy fact that $\varphi \in \mathrm{C}^\infty(\mathbb{R}^n \setminus \{0\})$, with decay at ∞. Then, φ extends to a $\mathrm{C}^\infty(\mathbb{R}^n)$ function with $\varphi(0) = 1$. This implies that $\varphi \in \mathscr{S}(\mathbb{R}^n)$ and so $\check\varphi \in \mathscr{S}(\mathbb{R}^n)$. Then, $\widehat{U_{\varepsilon,R}(a)}(\xi) = (-\varphi(R\xi) + \varphi(\varepsilon\xi))\hat a(\xi)$ so that

$$U_{\varepsilon,R} = -\check\varphi_R * a + \check\varphi_\varepsilon * a.$$

Then, it is enough to prove that $\varphi_R * a \to 0$ in H^1 as $R \to \infty$ and $\varphi_\varepsilon * a \to a$ in H^1 as $\varepsilon \to 0$.

Now, using $\int_{\mathbb{R}^n} a\, d\mathscr{L} = 0$, spt $a \subset B = B(x_B, r_B)$ and $\|a\|_1 \leq 1$, we leave it as an exercise to show that

$$|\varphi_R * a(x)| \leq \frac{C(n,\varphi)}{R}\left(\frac{1}{R^n}\left(1 + \frac{|x - x_B|}{R}\right)^{-n-1}\right)$$

when R is large,

$$\int_{\mathbb{R}^n} \varphi_R * a\, d\mathscr{L} = \int_{\mathbb{R}^n} \varphi_R\, d\mathscr{L} \int_{\mathbb{R}^n} a\, d\mathscr{L} = 0$$

and

$$\frac{R}{C(n,\varphi)} \varphi_R * a \in \mathrm{H}^1$$

with uniform norm with respect to R (ie., $\|\varphi_R * a\|_{\mathrm{H}^1} = O(\frac{1}{R})$).

Set $h_\varepsilon = \varphi_\varepsilon * a - a$. Then, we know that $h_\varepsilon \to 0$ in $\mathrm{L}^2(\mathbb{R}^n)$. Define functions $h_\varepsilon^1 = (h_\varepsilon - \mathrm{m}_{2B}\, h_\varepsilon)\chi_{2B}$ and h_ε^2 by $h_\varepsilon = h_\varepsilon^1 + h_\varepsilon^2$. Then, spt $h_\varepsilon^1 \subset 2B$,

$$\int_{2B} |h_\varepsilon^1|^2\, d\mathscr{L} \leq \int_{2B} |h_\varepsilon|^2\, d\mathscr{L} \to 0$$

and $\int_{2B} h_\varepsilon^1\, d\mathscr{L} = 0$. These facts then imply that $h_\varepsilon^1 \to 0$ in H^1 using the fact that $\{f \in \mathrm{L}^2(\mathbb{R}^n) : \mathrm{spt}\, f \subset 2B\}$ continuously embeds into H^1. Also,

$$\int_{\mathbb{R}^n} h_\varepsilon^2\, d\mathscr{L} = \int_{\mathbb{R}^n} h_\varepsilon\, d\mathscr{L} - \int_{\mathbb{R}^n} h_\varepsilon^1\, d\mathscr{L} = 0 - 0 = 0.$$

We leave it as an exercise to show that

$$\left|h_\varepsilon^2(x)\right| \leq C(n,\varphi)C(\varepsilon)(1 + |x - x_B|)^{-n-1}$$

for all $x \in \mathbb{R}^n$ and $\varepsilon < r_B$ where

$$C(\varepsilon) = \sup(\mathrm{m}_{2B}\, h_\varepsilon, \varepsilon).$$

With this fact in hand, $\|h_\varepsilon\|_{\mathrm{H}^1} \leq C(n,\varphi)C(\varepsilon) \to 0$.

(iii) We note that as before,

$$\langle \pi_b^{\mathrm{tr}}(f), g \rangle = \iint_{\mathbb{R}_+^{n+1}} Q_t(b)(x) P_t(g)(x) \tilde{Q}_t^{\mathrm{tr}}(f)(x) \, \frac{d\mathscr{L}(x)dt}{t}$$

for all $f, g \in \mathrm{L}^2(\mathbb{R}^n)$. We want to show that this holds when we have $f \in \mathrm{L}^\infty(\mathbb{R}^n)$ and $g \in \mathscr{A}^\infty$. As before, let $f_k = \chi_{[-2^k, 2^k]^n}$, and so

$$\begin{aligned}
\langle \pi_b^{\mathrm{tr}}(f), g \rangle &= \langle f, \pi_b(g) \rangle \\
&= \lim_{k \to \infty} \langle f_k, \pi_b(g) \rangle \\
&= \iint_{\mathbb{R}_+^{n+1}} Q_t(b)(x) P_t(g)(x) \tilde{Q}_t^{\mathrm{tr}}(f_k)(x) \, \frac{d\mathscr{L}(x)dt}{t}
\end{aligned}$$

Now, although $\tilde{Q}_t^{\mathrm{tr}}(f_k)(x) \to \tilde{Q}_t^{\mathrm{tr}}(f)(x)$ for all (x, t) we cannot apply Dominated convergence theorem since $\int_{\mathbb{R}^n} g \, d\mathscr{L} \neq 0$. Instead, cover $\mathbb{R}_+^{n+1}$ by set C_j as in the proof of Theorem 8.2.5.

For $j > 0$, we have the same estimates of $P_t^{\mathrm{tr}}(g)$ as for for $\tilde{Q}_t^{\mathrm{tr}}(g)$ in (iv). So, $\left| \tilde{Q}_t^{\mathrm{tr}}(g) \right| \leq C$ uniformly with respect to t, x, k. Then, take the limit.

For the situation when $j = 0$,

$$\left| \iint_{C_0} Q_t(b)(x) P_t(g)(x) \tilde{Q}_t^{\mathrm{tr}}(f - f_k)(x) \, \frac{d\mathscr{L}(x)dt}{t} \right| \leq \left(\iint_{C_0} \left| \tilde{Q}_t^{\mathrm{tr}}(f - f_k)(x) \right|^2 \frac{d\mathscr{L}(x)dt}{t} \right)^{\frac{1}{2}}$$
$$\left(\iint_{C_0} |Q_t(b)(x) P_t(g)(x)|^2 \, \frac{d\mathscr{L}(x)dt}{t} \right)^{\frac{1}{2}}$$

If k is large, $f - f_k = 0$ on a neighbourhood of $2B$. The region C_0 is above $2B$, so we can apply decay estimates to prove

$$\iint_{C_0} \left| \tilde{Q}_t^{\mathrm{tr}}(f - f_k)(x) \right|^2 \frac{d\mathscr{L}(x)dt}{t} \to 0.$$

The other term is dealt with by noting that $|Q_t(b)(x)|^2 \frac{d\mathscr{L}(x)dt}{t}$ is a Carleson measure.

Now, put $f = 1$, $\tilde{Q}_t^{\mathrm{tr}}(1)(x) = 0$ for all $(x, t) \in \mathbb{R}_+^{n+1}$ since $\int_{\mathbb{R}^n} \tilde{\psi} \, d\mathscr{L} = 0$. Thus, $\langle \pi_b^{\mathrm{tr}}(1), g \rangle = 0$ for all $g \in \mathscr{A}^\infty$ and so $\pi_b^{\mathrm{tr}}(1) = 0$ in BMO.

$\square$

Chapter 9

Littlewood-Paley Estimates

Definition 9.0.1 (Littlewood-Paley Estimate). *Let $(\mathcal{R}_t)_{t>0}$ be a family of operators on the Hilbert space $L^2(\mathbb{R}^n)$. Suppose there exists a $C > 0$ such that*

$$\int_0^\infty \|\mathcal{R}_t f\|_2^2 \, \frac{dt}{t} \le C\|f\|_2^2. \tag{LPE}$$

Such an estimate is called a Littlewood-Paley Estimate.

Example 9.0.2. *Let $\mathcal{R}_t = \psi_t \ast$ where $\psi \in \mathscr{S}(\mathbb{R}^n)$ with $\int_{\mathbb{R}^n} \psi \, d\mathscr{L} = 0$. Then,*

$$\int_0^\infty \|\mathcal{R}_t f\|_2^2 \frac{dt}{t} = C \int_{\mathbb{R}^n} \left(\int_0^\infty \left|\hat{\psi}(t\xi)\right|^2 \frac{dt}{t} \right) \left|\hat{f}(\xi)\right|^2 \, d\mathscr{L}(\xi) \le C(\psi)\|f\|_2^2.$$

Remark 9.0.3. *Unlike in the example, we will not have the luxury of the Fourier Transform in the general theory. This is our goal.*

Definition 9.0.4 (Operator ε-family). *Let $\varepsilon > 0$ and $(\mathcal{R}_t)_{t>0}$ be a family of Operators with kernels $K_t(x,y)$ where the map $(t,x,y) \mapsto K_t(x,y)$ is measurable satisfying:*

(i) For all $(t,x,y,z) \in \mathbb{R}_+ \times \mathbb{R}^{3n}$,

$$|K_t(x,y)| \le \frac{C}{t^n} \left(1 + \frac{|x-y|}{t} \right)^{-n-\varepsilon},$$

(ii) For $\delta > 0$,

$$|K_t(x,y) - K_t(x,z)| \le C \left(\frac{|y-z|}{t} \right)^\delta \frac{1}{t^n} \left(1 + \frac{|x-y|}{t} \right)^{-n-\varepsilon-\delta},$$

where $C < \infty$. Such a family of operators is called an ε-family.

Theorem 9.0.5 (Christ, Journé, Coifman, Meyer). *Let $(\mathcal{R}_t)_{t>0}$ be an ε-family. Then,*

(i) (LPE) holds,

(ii) The measure

$$|\mathcal{R}_t(1)(x)|^2 \, \frac{d\mathscr{L}(x)dt}{t}$$

is a Carleson measure,

are equivalent statements.

Remark 9.0.6. *The condition in (ii) only involves one test function, namely the constant function 1.*

The following proof of 9.0.5 is due to Fefferman-Stein.

Proof of 9.0.5 (i) $\implies$ *(ii).* Fix a Carleson box $R = B \times (0, r]$ where $r = \operatorname{rad} B$ and let $f \in L^\infty(\mathbb{R}^n)$. Let $f_1 = f\chi_{2B}$ and $f_2 = f\chi_{c2B}$ so that $f = f_1 + f_2$. Then,

$$\iint_R |\mathcal{R}_t(f_1)(x)|^2 \, \frac{d\mathscr{L}(x)dt}{t} \le C\|f_1\|_2^2 \le C\|f\|_\infty^2 \mathscr{L}(2B).$$

By noting that whenever $(x,t) \in R$ implies $x \in B$, we have

$$|\mathcal{R}_t(f_2)(x)| = \left| \int_{y \in {}^c 2B} K_t(x,y)f(y) \, d\mathscr{L}(y) \right|$$

$$\le \|f\|_\infty \int_{y \in {}^c 2B} \frac{C}{t^n} \left(1 + \frac{|x-y|}{t} \right)^{-n-\varepsilon} d\mathscr{L}(y) \le C\|f\|_\infty \left(\frac{t}{r} \right)^\varepsilon$$

Therefore,

$$\iint_R |\mathcal{R}_t(f_2)(x)|^2 \, \frac{d\mathscr{L}(x)dt}{t} \le C\|f\|_\infty^2 \iint_R \left(\frac{t}{r} \right)^{2\varepsilon} \frac{d\mathscr{L}(x)dt}{t} \le C\|f\|_\infty^2 \mathscr{L}(B)\frac{1}{2\varepsilon}.$$

$\square$

Before we proceed to prove the converse, we require the following important Lemma.

Lemma 9.0.7. *Let $(V_t)_{t>0}$ be an ε-family. Suppose that $V_t(1) = 0$. Then, there exist an $\varepsilon' \in (0, \varepsilon)$ and a $C > 0$ such that for all $(t, s) \in (0, \infty)^2$,*

$$\|V_t V_s^*\|_{\mathcal{L}(L^2(\mathbb{R}^n))} \le Ch\left(\frac{s}{t}\right)$$

where $h(x) = \inf\left\{x, \frac{1}{x}\right\}^{\varepsilon'}$.

Proof of 9.0.5 (ii) $\implies$ *(i).* First, define $V_t = \mathcal{R}_t - \mathcal{R}_t(1)P_t$ where $P_t = \varphi_t *$, $\varphi \in C_c^\infty(\mathbb{R}^n)$ with $\int_{\mathbb{R}^n} \varphi \, d\mathscr{L} = 1$. Then,

$$(\mathcal{R}_t(1)P_t)f(x) = \mathcal{R}_t(1)(x)P_t(f)(x) = \int_{\mathbb{R}^n} \mathcal{R}_t(1)(x)\varphi_t(x-y)f(y) \, d\mathscr{L}(y)$$

which shows that $(V_t)_{t>0}$ is an ε-family with the same ε as $\mathcal{R}_t$. Now, take $f \in \mathrm{L}^2(\mathbb{R}^n)$. Then,

$$\iint_{\mathbb{R}^{n+1}_+} |\mathcal{R}_t(1)(x)P_t(f)(x)|^2 \, \frac{d\mathscr{L}(x)dt}{t} \le C\|\mu\|_{\mathscr{C}}\|f\|_2^2$$

where

$$\mu = |\mathcal{R}_t(1)(x)|^2 \, \frac{d\mathscr{L}(x)dt}{t}.$$

Hence, (ii) implies that $|\mathcal{R}_t(1)(x)P_t(1)(x)|^2 \frac{d\mathscr{L}(x)dt}{t}$ satisfies (i) and so proving (i) for $\mathcal{R}_t$ reduces to proving (i) for V_t.

Firstly, we note that $V_t(1)(x) = 0$ for all $(x,t) \in \mathbb{R}^{n+1}_+$. We use a technique developed for a different purpose called the Cotlar-Knapp-Stein inequality or T^*T-argument. We have

$$\iint_{\mathbb{R}^{n+1}_+} |V_t(f)(x)|^2 \, \frac{d\mathscr{L}(x)dt}{t} = \int_0^\infty \langle V_t f, V_t f \rangle \, \frac{dt}{t} = \int_0^\infty \langle V_t^* V_t f, f \rangle \frac{dt}{t}$$

and for each t, $V_t^* V_t$ is a bounded (from ε-family hypothesis) self-adjoint operator on $\mathrm{L}^2(\mathbb{R}^n)$. We want the right hand side to be equal to $\langle Sf, f \rangle$ for some bounded operator S.

Fix r, R such that $0 < r \le R < \infty$, and let

$$S_{r,R} = \int_r^R V_t^* V_t \frac{dt}{t}$$

and it is a bounded operator on $\mathrm{L}^2(\mathbb{R}^n)$ since by the ε-family hypothesis $\|V_t^* V_t\|_{\mathcal{L}(\mathrm{L}^2(\mathbb{R}^n))}$ is uniformly bounded in t and so

$$\|S_{r,R}\|_{\mathcal{L}(\mathrm{L}^2(\mathbb{R}^n))} \le \int_r^R \|V_t^* V_t\|_{\mathcal{L}(\mathrm{L}^2(\mathbb{R}^n))} \frac{dt}{t} \le C \ln\left(\frac{R}{r}\right).$$

Also, $S_{r,R}$ is self adjoint and operator theory tells us that

$$\|S_{r,R}\|_{\mathcal{L}(\mathrm{L}^2(\mathbb{R}^n))}^m = \|S_{r,R}^m\|_{\mathcal{L}(\mathrm{L}^2(\mathbb{R}^n))}.$$

We write out $S_{r,R}^m$:

$$S_{r,R}^m = \int_r^R \cdots \int_r^R V_{t_1}^*(V_{t_1} V_{t_2}^*)V_{t_2} \cdots (V_{t_{m-1}} V_{t_m}^*)V_{t_m} \, \frac{dt_1}{t_1} \cdots \frac{dt_m}{t_m}.$$

and on application of Lemma 9.0.7, we get

$$\|S_{r,R}^m\|_{\mathcal{L}(\mathrm{L}^2(\mathbb{R}^n))}$$
$$\le \int_r^R \cdots \int_r^R \|V_{t_1}^*\|_{\mathcal{L}(\mathrm{L}^2(\mathbb{R}^n))} Ch\left(\frac{t_1}{t_2}\right) \cdots Ch\left(\frac{t_{m-1}}{t_m}\right) \|V_{t_m}\|_{\mathcal{L}(\mathrm{L}^2(\mathbb{R}^n))} \, \frac{dt_1}{t_1} \cdots \frac{dt_m}{t_m}$$
$$\le C^{m-1} \int_r^R \cdots \int_r^R \|V_{t_1}^*\|_{\mathcal{L}(\mathrm{L}^2(\mathbb{R}^n))} h\left(\frac{t_1}{t_2}\right) \cdots h\left(\frac{t_{m-1}}{t_m}\right) \|V_{t_m}\|_{\mathcal{L}(\mathrm{L}^2(\mathbb{R}^n))} \, \frac{dt_1}{t_1} \cdots \frac{dt_m}{t_m}.$$

Using that $\|V_{t_j}^*\|_{\mathcal{L}(\mathrm{L}^2(\mathbb{R}^n))}, \|V_{t_m}\|_{\mathcal{L}(\mathrm{L}^2(\mathbb{R}^n))} \in \mathrm{L}^\infty(\mathbb{R}_+)$, and $h \in \mathrm{L}^1(\mathbb{R}_+, \frac{dt}{t})$ with $\|h\|_1 = \frac{2}{\varepsilon'}$, we have integrating first in t_1, then t_2 and so on until t_{m-1}

$$\|S_{r,R}^m\|_{\mathcal{L}(\mathrm{L}^2(\mathbb{R}^n))} \le C^{m-1} \|h\|_1^{m-1} A^2 \int_r^R \frac{dt_m}{t_m} = C^{m-1} \|h\|_1^{m-1} A^2 \ln\left(\frac{R}{r}\right)$$

where $A = \sup_{t>0} \|V_t^*\|_{\mathcal{L}(L^2(\mathbb{R}^n))} = \sup_{t>0} \|V_t\|_{\mathcal{L}(L^2(\mathbb{R}^n))}$ and

$$\|S_{r,R}\|_{\mathcal{L}(L^2(\mathbb{R}^n))} \le \lim_{m\to\infty} \left(C^{m-1} \|h\|_1^{m-1} A^2 \ln\left(\frac{R}{r}\right) \right)^{\frac{1}{m}} = C\|h\|_1.$$

Then, we find that

$$\sup_{0<r<R<\infty} \left\| \int_r^R V_t^* V_t \frac{dt}{t} \right\|_{\mathcal{L}(L^2(\mathbb{R}^n))} \le C\|h\|_1$$

which proves

$$\int_0^\infty \|V_t f\|_2^2 \frac{dt}{t} = \lim_{r\to 0,\ R\to\infty} \int_r^R \langle V_t^* V_t f, f\rangle \frac{dt}{t} \le C\|h\|_1 \|f\|_2^2$$

for all $f \in L^2(\mathbb{R}^n)$. $\qquad\square$

Now, we return to the proof of the Lemma.

Proof of Lemma 9.0.7. We study $V_t V_s^*$ and its kernel. Let $V_t(x,y)$ and $V_s^*(x,y)$ denote the kernels of V_t and V_s^* respectively and

$$V_t V_s^*(f)(x) = \int_{\mathbb{R}^n} V_t(x,z) \int_{\mathbb{R}^n} V_s^*(z,y) f(y)\, d\mathcal{L}(y) d\mathcal{L}(z) = \int_{\mathbb{R}^n} K_{s,t}(x,y) f(y)\, d\mathcal{L}(y)$$

where

$$K_{s,t}(x,y) = \int_{\mathbb{R}^n} V_t(x,z) \overline{V_s(y,z)}\, d\mathcal{L}(z).$$

We note that by Fubini, this expression holds for all $f \in \cup_{1\le p\le\infty} L^p(\mathbb{R}^n)$.

Now, assume that $0 < s < t$. Then we have the estimates

$$|V_t(x,z)| \le \frac{C}{t^n}\left(1 + \frac{|x-z|}{t}\right)^{-n-\varepsilon} \quad \text{and} \quad |V_s(x,z)| \le \frac{C}{s^n}\left(1 + \frac{|x-z|}{s}\right)^{-n-\varepsilon}.$$

We leave it as an exercise to prove

$$\int_{\mathbb{R}^n} \frac{C}{t^n}\left(1 + \frac{|x-z|}{t}\right)^{-n-\varepsilon} \frac{C}{s^n}\left(1 + \frac{|y-z|}{s}\right)^{-n-\varepsilon} d\mathcal{L}(z)$$

$$\le C(n,\varepsilon)\frac{C}{t^n}\left(1 + \frac{|x-y|}{t}\right)^{-n-\varepsilon} \quad (\dagger)$$

for $0 < s \le t$.

Now, using oscillation of $y \mapsto V_s(y,z)$, we can write

$$K_{s,t}(x,y) = \int_{\mathbb{R}^n} (V_t(x,z) - V_t(x,y))\, \overline{V_s(y,z)}\, d\mathcal{L}(z)$$

and hence

$$|K_{s,t}(x,y)| \le \int_{\mathbb{R}^n} |V_t(x,z) - V_t(x,y)|\, |V_s(y,z)|\, d\mathcal{L}(z).$$

99

We also have

$$|V_t(x,z) - V_t(x,y)| \leq \begin{cases} \frac{C}{t^n} & \text{(decay)} \\ \frac{C}{t^n}\left(\frac{|y-z|}{t}\right)^\delta & \text{(regularity in } y\text{)} \end{cases}$$

and this implies that for all $\delta' \in (0, \delta)$, and some $C' > 0$,

$$|V_t(x,z) - V_t(x,y)| \leq \frac{C'}{t^n}\left(\frac{|y-z|}{t}\right)^{\delta'}.$$

Therefore, by choosing $0 < \delta' < \varepsilon$,

$$|K_{s,t}(x,y)| \leq \frac{C'}{t^n} \int_{\mathbb{R}^n} \left(\frac{|y-z|}{t}\right)^{\delta'} \left(1 + \frac{|y-z|}{s}\right)^{-n-\varepsilon} \frac{d\mathscr{L}(z)}{s^n} \leq \frac{C\, s^{\delta'}}{t^n\, t^{\delta'}}.$$

The technical estimate (†) gives us

$$|K_{s,t}(x,y)| \leq \frac{C}{t^n}\left(1 + \frac{|x-y|}{t}\right)^{-n-\varepsilon}.$$

Now, if $X \leq A$ and $X \leq B$, then for all $\theta \in [0,1]$ the estimate $X \leq A^\theta B^{1-\theta}$ holds. We apply this to $|K_{s,t}(x,y)|$ with θ such that $(1-\theta)(n+\varepsilon) = n + \nu > n$ to obtain

$$|K_{s,t}(x,y)| \leq \frac{C}{t^n}\left(\frac{s}{t}\right)^{\delta'\theta}\left(1 + \frac{|x-y|}{t}\right)^{-n-\nu}$$

for all $0 < s \leq t$. By symmetry, when $t \leq s$,

$$|K_{s,t}(x,y)| \leq \frac{C}{s^n}\left(\frac{t}{s}\right)^{\delta'\theta}\left(1 + \frac{|x-y|}{s}\right)^{-n-\nu}.$$

Using Young's inequality that $L^2 * L^1 \subset L^2$, we obtain

$$\|V_t V_s^*\|_{\mathcal{L}(L^2(\mathbb{R}^n))} \leq C \inf\left(\frac{s}{t}, \frac{t}{s}\right)^{\delta'\theta}.$$

$\square$

Chapter 10

$T(1)$ Theorem for Singular Integrals

This chapter concerns itself with the question of proving the boundedness of $L^2(\mathbb{R}^n)$ operators that are associated to Calderón-Zygmund kernels.

Definition 10.0.1 (Schwartz kernel). *Let $T : C_c^\infty(\mathbb{R}^n) \to C_c^\infty(\mathbb{R}^n)'$ be linear and continuous. Then, the uniquely given $K \in C_c^\infty(\mathbb{R}^{2n})'$ defined by*

$$\langle K, g \otimes f \rangle = \langle Tf, g \rangle$$

is called the Schwartz kernel of T.

We begin with the following definition.

Definition 10.0.2 (Singular integral operator). *Let $T : C_c^\infty(\mathbb{R}^n) \to C_c^\infty(\mathbb{R}^n)'$ be linear and continuous. Then T is said to be a Singular integral operator if its Schwartz kernel when restricted to $^c\Delta$ is a CZK_α for some $\alpha > 0$. We write $T \in \mathrm{SIO}$.*

We emphasise that while a Singular integral operator has CZK kernel, it is not a Calderón-Zygmund operator. Recall that a Calderón-Zygmund operator is bounded on $L^2(\mathbb{R}^n)$. The following examples emphasise that a Singular integral operator need not be bounded on $L^2(\mathbb{R}^n)$.

Example 10.0.3.   (i) $Tf(x) = |x|^2 f(x)$ *when $x \in \mathbb{R}^n$ is not bounded. But T is an SIO and the Schwartz kernel is given by $K(x, y) = |x|^2 \delta_0(x - y)$ and $K|_{^c\Delta} = 0 \in \mathrm{CZK}_\alpha$.*

   (ii) $Tf = f'$ *whenever $f \in C_c^\infty(\mathbb{R}^n)$ is not bounded on $L^2(\mathbb{R}^n)$. The Schwartz kernel here is given by $K(x, y) = -\delta_0'(x - y)$. Again, $K|_{^c\Delta} = 0 \in \mathrm{CZK}$*

   (iii) $Tf(x) = \ln|x| f(x)$ *when f is measurable. Then, $T|_{C_c^\infty(\mathbb{R}^n)} \in \mathrm{SIO}$ but $\ln|x| \notin L^\infty(\mathbb{R}^n)$. So, it is not bounded on $L^2(\mathbb{R}^n)$. It will be useful to emphasise for the sequel that that $T(1) = \ln|x| \in \mathrm{BMO}$.*

The previous examples illustrated that one of the problem for the boundedness of a $T \in \mathrm{SIO}$ on $L^2(\mathbb{R}^n)$ is the behaviour of the Schwartz kernel on the diagonal Δ. We will prove a theorem that $T \in \mathrm{SIO}$ is bounded in $\mathrm{L}^2(\mathbb{R}^n)$ if and only if both $T(1), T^{\mathrm{tr}}(1) \in \mathrm{BMO}$ and when T has the "weak boundedness property." In the sequel, we will give a rigorous formulation of this property. This is the key property that will give control on the diagonal of the Schwartz kernel of T.

Definition 10.0.4 ($\mathcal{N}_{B,q}$). *For $q \in \mathbb{N}$ and B a ball in $\mathbb{R}^n$ and any $\varphi \in \mathrm{C}_c^\infty(\mathbb{R}^n)$ with spt $\varphi \subset B$, define*

$$\mathcal{N}_{B,q}(\varphi) = \sup_{\alpha \in \mathbb{N}^n, \ |\alpha| \leq q} \left(\|\partial^\alpha \varphi\|_\infty (\mathrm{rad}\, B)^{|\alpha|} \right).$$

Remark 10.0.5. *This is a scale invariant quantity. Let* spt $\varphi \subset B(0,1)$ *and*

$$\psi(x) = \varphi\left(\frac{x - x_0}{r}\right)$$

for some $x_0 \in \mathbb{R}^n$ and $r > 0$. Then, spt $\psi \subset B(x_0, r)$ *and* $\mathcal{N}_{B(x_0,r),q}(\psi) = \mathcal{N}_{B(0,1),q}(\varphi)$.

Proposition 10.0.6. *Let $T \in \mathrm{SIO}$ and fix $x_0 \in \mathbb{R}^n$, $r > 0$ and define $T_{x_0,r}$ by*

$$\langle T_{x_0,r}\varphi, \psi \rangle = \langle T\varphi\left(\frac{\cdot - x_0}{r}\right), \psi\left(\frac{\cdot - x_0}{r}\right)\rangle.$$

Then, $T_{x_0,r} \in \mathrm{SIO}$ and its kernel is given by $K_{x_0,r}(x,y) = r^{2n} K(rx + x_0, ry + x_0)$.

Proof. Exercise. $\qquad\qquad\qquad\qquad\qquad\qquad\qquad\qquad\qquad\qquad\qquad\qquad\square$

Corollary 10.0.7. *$T \in \mathcal{L}(\mathrm{L}^2(\mathbb{R}^n))$ if and only if $T_{x_0,r} \in \mathcal{L}(\mathrm{L}^2(\mathbb{R}^n))$ and $\|T_{x_0,r}\|_{\mathcal{L}(\mathrm{L}^2(\mathbb{R}^n))} \leq r^n \|T\|_{\mathcal{L}(\mathrm{L}^2(\mathbb{R}^n))}$.*

Definition 10.0.8 (Weak boundedness property). *Let $T \in \mathrm{SIO}$. We say that T has the weak boundedness property (or WBP) if there exists a $q \in \mathbb{N}$, $q > 0$, and a $C > 0$ such that for all balls B and all $\varphi, \psi \in \mathrm{C}_c^\infty(B)$,*

$$|\langle T\varphi, \psi \rangle| \leq C \mathscr{L}(B) \mathcal{N}_{B,q}(\varphi) \mathcal{N}_{B,q}(\psi).$$

Remark 10.0.9. *We note that $\|\varphi\|_2 \leq \mathscr{L}(B)^{\frac{1}{2}} \|\varphi\|_\infty \leq \mathscr{L}(B)^{\frac{1}{2}} \mathcal{N}_{B,q}(\varphi)$. Therefore, if $T \in \mathcal{L}(\mathrm{L}^2(\mathbb{R}^n))$, then*

$$|\langle T\varphi, \psi \rangle| \leq \|T\|_{\mathcal{L}(\mathrm{L}^2(\mathbb{R}^n))} \|\varphi\|_2 \|\psi\|_2 \leq \|T\|_{\mathcal{L}(\mathrm{L}^2(\mathbb{R}^n))} \mathscr{L}(B) \mathcal{N}_{B,q}(\varphi) \mathcal{N}_{B,q}(\psi).$$

We now deal with the issue of giving meaning to T acting on the constant function $f \equiv 1$. Note that this is non trivial since T is an operator defined on compactly supported functions.

Definition 10.0.10. *Define*

$$\mathscr{D}_0(\mathbb{R}^n) = \left\{ \psi \in \mathrm{C}_c^\infty(\mathbb{R}^n) : \int_{\mathbb{R}^n} \psi \, d\mathscr{L} = 0 \right\}.$$

Lemma 10.0.11. *Let $T \in \mathrm{SIO}$ and $\psi \in \mathscr{D}_0(\mathbb{R}^n)$. Let B be a ball such that* spt $\psi \subset B$. *Then, $T\psi \in \mathrm{L}^1({}^c\overline{2B})$ and*

$$\|T\psi\|_{\mathrm{L}^1({}^c\overline{2B})} \leq \|K\|_{\mathrm{CZK}_\alpha} C(n, \alpha) \|\psi\|_{\mathrm{L}^1(B)}.$$

Remark 10.0.12. *In writing $T\psi \in \mathrm{L}^1({}^c\overline{2B})$, we mean the distribution $T\psi|_{{}^c\overline{2B}} \in \mathrm{L}^1({}^c\overline{2B})$.*

Proof. Fix $\varphi \in \mathrm{C}_c^\infty(\mathbb{R}^n)$ and spt $\varphi \subset {}^c\overline{2B}$. Then,

$$\langle T\psi, \varphi \rangle = \langle K, \varphi \otimes \psi \rangle = \langle K|_{{}^c\Delta}, \varphi \otimes \psi \rangle$$

since spt $\varphi \otimes \psi \subset {}^c\Delta$. Certainly,

$$\langle K|_{{}^c\Delta}, \varphi \otimes \psi \rangle = \iint_{{}^c\Delta} K(x,y)\varphi(x)\psi(y)\, d\mathscr{L}(y) d\mathscr{L}(x) = \int_{\mathbb{R}^n} \left(\int_{\mathbb{R}^n} K(x,y)\psi(y)\, d\mathscr{L}(y) \right) \varphi(x)\, d\mathscr{L}(x)$$

and so $T\psi|_{{}^c\overline{2B}}$ agrees with

$$\int_{\mathbb{R}^n} K(x,y)\psi(y)\, d\mathscr{L}(y).$$

Now, since $\int_{\mathbb{R}^n} \psi\, d\mathscr{L} = 0$,

$$\int_{\mathbb{R}^n} K(x,y)\psi(y)\, d\mathscr{L}(y) = \int_{\mathbb{R}^n} (K(x,y) - K(x,y_B))\,\psi(y)\, d\mathscr{L}(y)$$

where y_B is the centre of B and

$$\left| \int_{\mathbb{R}^n} K(x,y)\psi(y)\, d\mathscr{L}(y) \right| \leq \|K\|_{\mathrm{CZK}_\alpha} \int_{\mathbb{R}^n} \frac{(\mathrm{rad}\, B)^\alpha}{|x - y_B|^{n+\alpha}} |\psi(y)|\, d\mathscr{L}(y)$$

$$\leq \|K\|_{\mathrm{CZK}_\alpha} \frac{(\mathrm{rad}\, B)^\alpha}{|x - y_B|^{n+\alpha}} \|\psi\|_{\mathrm{L}^1(B)}.$$

The conclusion is achieved by integrating over $x \in {}^c\overline{2B}$. $\qquad\square$

Proposition 10.0.13. *Let $f \in \mathrm{C}^\infty \cap \mathrm{L}^\infty(\mathbb{R}^n)$ and $\psi \in \mathscr{D}_0(\mathbb{R}^n)$ and let B be a ball such that* spt $\psi \subset B$. *Then, whenever $\eta \in \mathrm{C}_c^\infty(\mathbb{R}^n)$ such that* spt $\eta \subset 4B$ *and $\eta \equiv 1$ on $3B$,*

$$\langle T(f\eta), \psi \rangle + \int_{\mathbb{R}^n} (1 - \eta)(x)f(x)T^{\mathrm{tr}}(\psi)(x)\, d\mathscr{L}(x)$$

is well defined and does not depend on the choice of η.

Proof. We note that ψ and $f\eta \in \mathrm{C}_c^\infty(\mathbb{R}^n)$ and this implies that $\langle T(f\eta), \psi \rangle$ exists. By the application of Lemma 10.0.11, we have $T^{\mathrm{tr}}(\psi)|_{{}^c\overline{2B}} \in \mathrm{L}^1({}^c\overline{2B})$. Also, $(1 - \eta)f \in \mathrm{L}^\infty(\mathbb{R}^n)$ and spt $(1 - \eta)f \subset {}^c\overline{2B}$.

To show the independence of η, choose η_1 and η_2 with the desired properties. Then, we note that

$$\langle T(f\eta_1), \psi \rangle + \int_{\mathbb{R}^n} (1 - \eta_1)(x)f(x)T^{\mathrm{tr}}(\psi)(x)\, d\mathscr{L}(x)$$

$$= \langle T(f\eta_2), \psi \rangle + \int_{\mathbb{R}^n} (1 - \eta_2)(x)f(x)T^{\mathrm{tr}}(\psi)(x)\, d\mathscr{L}(x)$$

is equivalent to

$$\langle T(f(\eta_1 - \eta_2)), \psi \rangle = \int_{\mathbb{R}^n} (\eta_1 - \eta_2)(x)f(x)T^{\mathrm{tr}}(\psi)(x)\, d\mathscr{L}(x).$$

Noting that spt $\psi \subset B$ and spt $(\eta_1 - \eta_2)f \subset \overline{4B \setminus 3B}$, we compute

$$\langle T(f(\eta_1 - \eta_2)), \psi \rangle = \iint_{\mathbb{R}^n} K(x,y)f(y)(\eta_1 - \eta_2)(y)\psi(x) \, d\mathscr{L}(x)d\mathscr{L}(y)$$

$$= \int_{\mathbb{R}^n} (\eta_1 - \eta_2)(x)f(x) \left(\int_{\mathbb{R}^n} K(y,x)\psi(y) \, d\mathscr{L}(y) \right) d\mathscr{L}(x)$$

by the application of Fubini. $\qquad\square$

As a consequence, we are able to make the following definition.

Definition 10.0.14. *Whenever $f \in C^\infty \cap L^\infty(\mathbb{R}^n)$, define Tf by*

$$\langle Tf, \psi \rangle = \langle T(f\eta), \psi \rangle + \int_{\mathbb{R}^n} (1-\eta)(x)f(x)T^{\mathrm{tr}}(\psi)(x) \, d\mathscr{L}(x)$$

whenever $\psi \in \mathscr{D}_0(\mathbb{R}^n)$ and $\eta \in C_c^\infty(\mathbb{R}^n)$ such that spt $\psi \subset B$, spt $\eta \subset 4B$ and $\eta \equiv 1$ on $3B$.

Remark 10.0.15. *If $f \in C_c^\infty(\mathbb{R}^n)$ and $\eta = 1$ on spt f, the right hand side agrees with $\langle T(f\eta), \psi \rangle$ which is $\langle Tf, \psi \rangle$. Hence, this definition is an extension of the original one.*

Proposition 10.0.16. *Whenever $f \in C^\infty \cap L^\infty(\mathbb{R}^n)$, we have $Tf \in \mathscr{D}_0(\mathbb{R}^n)'$.*

Proof. We want to show that $Tf \in \mathscr{D}_0(\mathbb{R}^n)'$ with bounds depending only on f. Let B be a ball and $\psi \in \mathscr{D}_0(\mathbb{R}^n)$ a such that spt $\psi \subset B$. Fix an η as in the definition of Tf.

As $T \in \mathcal{L}(C_c^\infty(\mathbb{R}^n), C_c^\infty(\mathbb{R}^n)')$, there exists an integer $M > 0$ such that

$$|\langle T(f\eta), \psi \rangle| \leq C(M, 4B) \sup_{|\alpha| \leq M} \|\partial^\alpha(f\eta)\|_\infty \sup_{|\alpha| \leq M} \|\partial^\alpha \psi\|_\infty$$

$$\leq C'(M, 4B) \sup_{|\alpha| \leq M} \|\partial^\alpha f\|_{L^\infty(B)} \sup_{|\alpha| \leq M} \|\partial^\alpha \psi\|_\infty$$

Now,

$$\int_{\mathbb{R}^n} |(1-\eta)(x)f(x)T^{\mathrm{tr}}(\psi)(x)| \, d\mathscr{L}(x) \leq \int_{c\overline{2B}} \|1-\eta\|_\infty \|f\|_\infty |T^{\mathrm{tr}}(\psi)(x)| \, d\mathscr{L}(x)$$

$$\leq \|1-\eta\|_\infty \|f\|_\infty C(n,\alpha)\|K\|_{\mathrm{CZK}_\alpha}\|\psi\|_1$$

$$\leq C(n,\alpha)(1 + \|\eta\|_\infty)\|f\|_\infty \|K\|_{\mathrm{CZK}_\alpha}\|\psi\|_\infty \mathscr{L}(B).$$

Altogether, we have shown

$$|\langle Tf, \psi \rangle| \leq C(M, B, f) \sup_{|\alpha| \leq M} \|\partial^\alpha \psi\|_\infty.$$

This shows that $\psi \mapsto \langle Tf, \psi \rangle$ is continuous on $\mathscr{D}_0(\mathbb{R}^n)$. $\qquad\square$

We now present the crucial theorem of this chapter.

Theorem 10.0.17 ($T(1)$ *Theorem of David-Journé ('84))*. *Let $T \in$ SIO. Then, $T \in$* $\mathcal{L}(\mathrm{L}^2(\mathbb{R}^n))$ *if and only if T possesses the weak boundedness property, $T(1), T^{\mathrm{tr}}(1) \in$ BMO.*

Remark 10.0.18. *When we write $T \in \mathcal{L}(\mathrm{L}^2(\mathbb{R}^n))$, we mean T extends to such an operator since what we prove is that*

$$|\langle Tf, g \rangle| \le C \|f\|_2 \|g\|_2$$

whenever $f, g \in \mathrm{C}_c^\infty(\mathbb{R}^n)$.

The following Lemma is of use in the proof of the $T(1)$ Theorem. We leave its proof as an exercise.

Lemma 10.0.19. *Let $\varphi \in \mathrm{C}_c^\infty(\mathbb{R}^n)$ be a radial function such that $\int_{\mathbb{R}^n} \varphi \, d\mathscr{L} = 1$, and* spt $\varphi \subset B(0, \frac{1}{2})$. *Let $P_t = \varphi_t *$. Then for all $f \in \mathrm{C}_c^\infty(\mathbb{R}^n)$, $\lim_{t \to 0} P_t^2(f) = f$ in $\mathrm{C}_c^\infty(\mathbb{R}^n)$.*

Proof of the $T(1)$ Theorem. We already have one direction: if $T \in \mathcal{L}(\mathrm{L}^2(\mathbb{R}^n))$ we have already seen that T has WBP and $T(1), T^{\mathrm{tr}}(1) \in$ BMO. We prove the converse.

Let $T_0 = T - \pi_{T(1)} - \pi^{\mathrm{tr}}_{T^{\mathrm{tr}}(1)}$ where π_b is the paraproduct with $b \in$ BMO. We note we can choose π_b such that $\pi_b \in \mathrm{CZO}_1 \subset$ SIO, $\pi_b(1) = b$ in BMO and $\pi_b^{\mathrm{tr}} = 0$ in BMO. Therefore, $T_0 \in$ SIO and possesses WBP. Also, $T_0(1) = T(1) - T(1) - 0 = 0$ and $T_0^{\mathrm{tr}}(1) = T^{\mathrm{tr}}(1) - 0 - T^{\mathrm{tr}}(1) = 0$ in BMO. Since $T \in \mathcal{L}(\mathrm{L}^2(\mathbb{R}^n))$ if and only if $T_0 \in \mathcal{L}(\mathrm{L}^2(\mathbb{R}^n))$, we can assume that $T(1) = 0$ and $T^{\mathrm{tr}}(1) = 0$ in BMO.

Take a φ and P_t as in the hypothesis of Lemma 10.0.19. Fix $f, g \in \mathrm{C}_c^\infty(\mathbb{R}^n)$ and by the same Lemma,

$$\langle Tf, g \rangle = \lim_{\varepsilon \to 0} \langle TP_\varepsilon^2 f, P_\varepsilon^2 g \rangle - \lim_{R \to \infty} \langle TP_R^2 f, P_R^2 g \rangle = \lim_{\varepsilon \to 0} \left(\langle TP_\varepsilon^2 f, P_\varepsilon^2 g \rangle - \langle TP_{\frac{1}{\varepsilon}}^2 f, P_{\frac{1}{\varepsilon}}^2 g \rangle \right).$$

The limit when $R \to \infty$ exists thanks to the WBP property. Indeed, if f, g have support say in the ball $B(0, A)$ then $P_R^2 f, P_R^2 g$ have support in $B(0, R + A)$ and one can check that their $\mathcal{N}_{B,q}$ norms are bounded by R^{-n}. By WBP, $|\langle TP_R^2 f, P_R^2 g \rangle|$ is controlled by R^{-n}. Also, the map $t \in (0, \infty) \mapsto P_t^2 f \in \mathrm{C}_c^\infty(\mathbb{R}^n)$ is C^1 and therefore,

$$\langle Tf, g \rangle = \lim_{\varepsilon \to 0} \int_\varepsilon^{\frac{1}{\varepsilon}} \frac{d}{dt} \langle TP_t^2, P_t^2 g \rangle \, dt.$$

Here,

$$\frac{d}{dt} \langle TP_t^2, P_t^2 g \rangle \, dt = \langle T \left(\frac{d}{dt} P_t^2 f \right), P_t^2 g \rangle + \langle TP_t^2 f, \frac{d}{dt} P_t^2 g \rangle$$

and since $\langle TP_t^2 f, \frac{d}{dt} P_t^2 g \rangle = \langle T^{\mathrm{tr}} \left(\frac{d}{dt} P_t^2 g \right), P_t^2 f \rangle$, it is enough to just treat $\langle T \left(\frac{d}{dt} P_t^2 f \right), P_t^2 g \rangle$.

Now, we note that $P_t^2 f \in \mathrm{C}_c^\infty(\mathbb{R}^n) \subset \mathscr{S}(\mathbb{R}^n)$ and we consider the spatial Fourier Transform. On noting that φ is radial if and only if $\hat{\varphi}$ is radial, we have

$$\widehat{\frac{d}{dt} P_t^2 f}(\xi) = \frac{d}{dt} \widehat{P_t^2 f}(\xi) = \frac{d}{dt} \hat{\varphi}(t\xi)^2 \hat{f}(\xi) = 2 \sum_{k=1}^n \xi_k (\partial_k \hat{\varphi}(t\xi)) \hat{\varphi}(t\xi).$$

So define $\psi_k(x) = -\imath x_k \varphi(x)$ and $\tilde{\psi}_k(x) = \frac{1}{\imath}\partial_k\varphi(x)$ so that $\psi_k, \tilde{\psi}_k \in C_c^\infty(B(0,\frac{1}{2}))$ with $\int_{\mathbb{R}^n}\psi_k\,d\mathscr{L} = \int_{\mathbb{R}^n}\tilde{\psi}_k\,d\mathscr{L} = 0$. Also, $\hat{\psi}_k(\xi) = \partial_k\hat{\varphi}(\xi)$ and $\hat{\tilde{\psi}}_k(\xi) = \xi_k\hat{\varphi}(\xi)$ and therefore,

$$\widehat{\frac{d}{dt}P_t^2 f}(\xi) = \frac{2}{t}\sum_{k=1}^n t\left(\xi_k\partial_k\varphi(\hat{t}\xi)\right)\hat{\varphi}(t\xi)\hat{f}(\xi) = \frac{2}{t}\sum_{k=1}^n \hat{\psi}_k(t\xi)\hat{\tilde{\psi}}_k(t\xi)\hat{f}(\xi).$$

Letting $Q_{k,t} = \psi_{k,t}*$ and $\tilde{Q}_{k,t} = \tilde{\psi}_{k,t}*$ we have

$$\frac{d}{dt}P_t^2 f = \frac{2}{t}\sum_{k=1}^n Q_{k,t}\tilde{Q}_{k,t}f.$$

We consider $Q_{k,t}\tilde{Q}_{k,t}f$ for each k. So fix k and $\varepsilon > 0$. We prove that there exists a $C > 0$ (independent of ε) such that

$$\left|\int_\varepsilon^{\frac{1}{\varepsilon}}\langle TQ_{k,t}\tilde{Q}_{k,t}f, P_t^2 g\rangle\,\frac{dt}{t}\right| \le C\|f\|_2\|g\|_2.$$

Let

$$I_\varepsilon = \int_\varepsilon^{\frac{1}{\varepsilon}}\langle TQ_{k,t}\tilde{Q}_{k,t}f, P_t^2 g\rangle\,\frac{dt}{t} = \int_\varepsilon^{\frac{1}{\varepsilon}}\langle\tilde{Q}_{k,t}f, Q_{k,t}^{\mathrm{tr}}T^{\mathrm{tr}}P_t^2 g\rangle\,\frac{dt}{t},$$

and $R_{k,t} = Q_{k,t}^{\mathrm{tr}}T^{\mathrm{tr}}P_t^2$. Then, by Cauchy-Schwarz

$$|I_\varepsilon| \le \left(\int_\varepsilon^{\frac{1}{\varepsilon}}\|R_{k,t}g\|_2^2\,\frac{dt}{t}\right)^{\frac{1}{2}}\left(\int_\varepsilon^{\frac{1}{\varepsilon}}\|\tilde{Q}_{k,t}f\|_2^2\,\frac{dt}{t}\right)^{\frac{1}{2}}.$$

By an LPE estimate, we get that

$$\left(\int_\varepsilon^{\frac{1}{\varepsilon}}\|\tilde{Q}_{k,t}f\|_2^2\,\frac{dt}{t}\right)^{\frac{1}{2}} \le C(\psi_k)\|f\|_2$$

and so it is enough to show that

$$\int_0^\infty \|R_{k,t}g\|_2^2\,\frac{dt}{t} \le C(\varphi, \psi_k, T)\|g\|_2^2.$$

We examine the kernel of $R_{k,t}$. Pick a $h \in C_c^\infty(\mathbb{R}^n)$. First, we note that $\langle R_{k,t}g, h\rangle = \langle TQ_{k,t}h, P_t^2 g\rangle$. For $u, v \in \mathbb{R}^n$, set

$$\psi_{k,t}^u(x) = \psi_{k,t}(x-u) \quad \text{and} \quad \tilde{\varphi}_t^v(x) = \tilde{\varphi}_t(x-v), \quad \tilde{\varphi} = \varphi * \varphi,$$

so that

$$Q_{k,t}h = \int_{\mathbb{R}^n}\psi_{k,t}^u h(u)\,d\mathscr{L}(u) \quad \text{and} \quad P_t^2 g = \int_{\mathbb{R}^n}\tilde{\varphi}_t^v g(v)\,d\mathscr{L}(v).$$

Using the continuity of $T : C_c^\infty(\mathbb{R}^n) \to C_c^\infty(\mathbb{R}^n)'$, we can write

$$\langle TQ_{k,t}h, P_t^2 g\rangle = \int_{\mathbb{R}^n}\int_{\mathbb{R}^n} h(u)\langle T\psi_{k,t}^u, \tilde{\varphi}_t^v\rangle g(v)\,d\mathscr{L}(u)d\mathscr{L}(v)$$

106

and so the kernel of $R_{k,t}$ is $K_{k,t}(u,v) = \langle T\psi_{k,t}^u, \tilde{\varphi}_t^v \rangle$.

We estimate $K_{k,t}(u,v)$. First, suppose $|u - v| \leq 3t$. then, spt $\psi_{k,t}^u \subset B(u, \frac{t}{2})$ and spt $\tilde{\varphi}_t^v \subset B(v,t)$ and by setting $B = B(v, 4t)$ we have that spt $\psi_{k,t}^u$, spt $\tilde{\varphi}_t^v \subset B$. By the fact that T has WBP,

$$|K_{k,t}(u,v)| \leq C\mathscr{L}(B)\mathcal{N}_{B,q}(\psi_{k,t}^u)\mathcal{N}_{B,q}(\tilde{\varphi}_t^v)$$

and so

$$|K_{k,t}(u,v)| \leq C\frac{C(n, \psi_k, \varphi)}{t^n}.$$

Now, suppose $|u - v| \geq 3t$. As before, we have spt $\psi_{k,t}^u \subset B(u, \frac{t}{2})$ and also $\int_{\mathbb{R}^n} \psi_{k,t}^u \, d\mathscr{L} = 0$ which gives

$$\left|T\psi_{k,t}^u(x)\right| \leq C\|K\|_{\mathrm{CZK}_\alpha}\frac{t^\alpha}{|x - u|^{n+\alpha}}\|\psi_{k,t}^u\|_1$$

where K is the kernel of T and when $x \notin B(u,t)$. Now, spt $\tilde{\varphi}_t^v \subset B(v,t) \subset {}^cB(u,t)$ and so

$$\left|\langle T\psi_{k,t}^u, \tilde{\varphi}_t^v \rangle\right| = \int_{\mathbb{R}^n} T\psi_{k,t}^u(x)\varphi_t^v(x) \, d\mathscr{L}(x) \leq C\|K\|_{\mathrm{CZK}_\alpha}\frac{t^\alpha}{|v - u|^{n+\alpha}}\|\psi_k\|_1\|\tilde{\varphi}\|_1$$

since $\|\psi_{k,t}^u\|_1 = \|\psi_{k,t}\|_1 = \|\psi_k\|_1$ and $\|\tilde{\varphi}_t\|_1 = \|\tilde{\varphi}\|_1$.

Now, we consider the regularity of $K_{k,t}(u,v)$ with respect to v. We note that $v \mapsto \tilde{\varphi}_t^v \in C_c^\infty(\mathbb{R}^n)$ is a C^1 function and

$$\partial_{v_l}\tilde{\varphi}_t^v = -\frac{1}{t}(\partial_l\tilde{\varphi})_t^v \quad \text{and} \quad \partial_{v_l}K_{k,t}(u,v) = -\frac{1}{t}\langle T\psi_{k,t}^u, (\partial_l\tilde{\varphi})_t^v)\rangle.$$

This gives

$$|\partial_{v_l}K_{k,t}(u,v)| \leq \frac{C}{t^{n+1}}\left(1 + \frac{|u-v|}{t}\right)^{-n-\alpha}.$$

So $(R_t)_{t>0}$ is an ε'-family for some $\varepsilon' < \alpha$.

We compute $R_{k,t}(1)$. Choose $\eta \in C_c^\infty(\mathbb{R}^n)$ with $\eta \equiv 1$ on $B(u, \frac{3t}{2})$ and spt $\eta \subset B(u, 2t)$. Then,

$$R_{k,t}(1)(u) = \int_{\mathbb{R}^n} K_{k,t}(u,v) \, d\mathscr{L}(v)$$

$$= \int_{\mathbb{R}^n} \langle T\psi_{k,t}^u, \tilde{\varphi}_t^v \rangle \, d\mathscr{L}(v)$$

$$= \int_{\mathbb{R}^n} \langle T\psi_{k,t}^u, \eta\tilde{\varphi}_t^v \rangle \, d\mathscr{L}(v) + \int_{\mathbb{R}^n} \langle T\psi_{k,t}^u, (1-\eta)\tilde{\varphi}_t^v \rangle \, d\mathscr{L}(v)$$

$$= \langle T\psi_{k,t}^u, \int_{\mathbb{R}^n} \eta\tilde{\varphi}_t^v \, d\mathscr{L}(v)\rangle + \int_{\mathbb{R}^n} \left(\int_{\mathbb{R}^n} T\psi_{k,t}^u(x)(1-\eta)(x)\tilde{\varphi}_t^v(x) \, d\mathscr{L}(x)\right) d\mathscr{L}(v).$$

First, we have

$$\int_{\mathbb{R}^n} \eta\tilde{\varphi}_t^v \, d\mathscr{L}(v) = \eta$$

in $C_c^\infty(\mathbb{R}^n)$ so $\langle T\psi_{k,t}^u, \int_{\mathbb{R}^n} \eta\tilde{\varphi}_t^v \, d\mathscr{L}(v)\rangle = \langle T\psi_{k,t}^u, \eta\rangle$. Also, $T\psi_{k,t}^u \in L^1(\text{spt }(1-\eta))$ in x, $(1-\eta) \in L^\infty(\mathbb{R}^n)$ in x, $\tilde{\varphi}_t^v \in L^1(\mathbb{R}^n)$ in v and $\tilde{\varphi}_t^v \in L^\infty(\mathbb{R}^n)$ in x. By Fubini coupled with $\int_{\mathbb{R}^n} \tilde{\varphi}_{k,t}^v \, d\mathscr{L}(v) = 1$

$$\int_{\mathbb{R}^n} \left(\int_{\mathbb{R}^n} T\psi_{k,t}^u(x)(1-\eta)(x)\varphi_t^v(x) \, d\mathscr{L}(x)\right) d\mathscr{L}(v) = \int_{\mathbb{R}^n} T\psi_{k,t}^u(x)(1-\eta)(x) \, d\mathscr{L}(x).$$

By definition of $T^{\text{tr}}(1)$ and using $T^{\text{tr}}(1) = 0$ in BMO,

$$R_{k,t}(1)(u) = \int_{\mathbb{R}^n} \left(\int_{\mathbb{R}^n} T\psi_{k,t}^u(x)(1-\eta)(x)d\mathcal{L}(x) \right) + \langle T\psi_{k,t}^u, \eta \rangle = \langle T^{\text{tr}}(1), \psi_{k,t}^u \rangle = 0.$$

Combining these facts, we have shown that $R_{k,t}$ is an ε'-family with $R_{k,t}(1) = 0$ for all t and k. In particular

$$|R_{k,t}(1)(x)|^2 \, \frac{d\mathcal{L}(x)dt}{t}$$

is a Carleson measure. Therefore, by Theorem 9.0.5,

$$\int_0^\infty \|R_{k,t}g\|_2^2 \, \frac{dt}{t} \le C\|g\|_2^2$$

and the proof is complete. $\qquad\qquad\qquad\qquad\qquad\qquad\qquad\qquad\qquad\qquad\quad \square$

Remark 10.0.20 (Avoiding the reduction to $T(1) = T^{\text{tr}}(1) = 0$). *We note that for all $t > 0$ and all $u \in \mathbb{R}^n$ the equality $R_{k,t}(1)(u) = \langle T^{\text{tr}}(1), \psi_{k,t}^u \rangle$ holds when $T^{\text{tr}}(1) \in \text{BMO}$. Going back again to the definition of $\langle T^{\text{tr}}(1), \psi_{k,t}^u \rangle$ and comparing with the definition of $\check{Q}_{k,t}(T^{\text{tr}}(1))(u)$ when $T^{\text{tr}}(1) \in \text{BMO}$ (since $\check{Q}_{k,t} = \check{\psi}_{k,t}^u \, *$), we can see that*

$$\langle T^{\text{tr}}(1), \psi_{k,t}^u \rangle = \check{Q}_{k,t}(T^{\text{tr}}(1))(u).$$

Thus, $R_{k,t}(1) = \check{Q}_{k,t}(T^{\text{tr}}(1))$. As the Littlewood-Paley estimates combined with the decay of $\check{Q}_{k,t}$ and $T^{\text{tr}}(1) \in \text{BMO}$ imply that

$$\left| \check{Q}_{k,t}(T^{\text{tr}}(1))(x) \right|^2 \frac{d\mathcal{L}(x)dt}{t}$$

is a Carleson measure we can conclude by applying Theorem 9.0.5.

Bibliography

[Boa54] Ralph Boas, *Entire functions*, Elsevier, 1954.

[CR77] Weiss G. Coifman R., *Extensions of hardy spaces and their use in analysis*, Bull. Amer. Math. Soc. **83** (1977), no. 4, 569–645.

[Fed96] Herbert Federer, *Geometric measure theory*, Springer-Verlag Berlin, 1996.

[G.81] De Barra G., *Measure and integration*, New Age International Ltd., 1981.

[GC91] Evans L. Gariepy C., *Measure theory and fine properties of functions*, CRC Press Inc., 1991.

[Mey92] Yves Meyer, *Wavelets and operators*, Cambridge University Press, 1992.

[Ste71a] E.M. Stein, *Singular integrals and differentability properties of functions*, Princeton University Press, 1971.

[Ste71b] Weiss G. Stein, E.M., *Introduction to fourier analysis on euclidean spaces*, Princeton University Press, 1971.

[Ste93] E. M. Stein, *Harmonic analysis: real-variable methods, orthogonality, and oscillatory integrals*, Princeton Mathematical Series, vol. 43, Princeton University Press, Princeton, NJ, 1993.

[Yos95] Kôsaku Yosida, *Functional analysis*, Springer-Verlag, 1995.

Index

www.ingramcontent.com/pod-product-compliance
Lightning Source LLC
Chambersburg PA
CBHW042113210326
41518CB00043B/2600